Justice in Genetics

Justice in Genetics

Intellectual Property and Human Rights from a Cosmopolitan Liberal Perspective

Louise Bernier

University of Sherbrooke, Québec, Canada

Edward Elgar

Cheltenham, UK • Northampton, MA, USA

Published by
Edward Elgar Publishing Limited
The Lypiatts
15 Lansdown Road
Cheltenham
Glos GL50 2JA
UK

Edward Elgar Publishing, Inc.
William Pratt House
9 Dewey Court
Northampton
Massachusetts 01060
USA

A catalogue record for this book is available from the British Library

Library of Congress Control Number: 2009940646

ISBN 978 1 84844 315 0

Typeset by Cambrian Typesetters, Camberley, Surrey
Printed and bound by MPG Books Group, UK

Contents

Acronyms

AIDS	Acquired Immunodeficiency Syndrome
CBD	Convention on Biological Diversity
CF	Cystic Fibrosis
CIPP	Center for Intellectual Property Policy
EMBO	European Molecular Biology Organization
ESC	European Social Charter
ESTs	Expressed Sequence Tags
FAO	Food and Agriculture Organization
FTAs	Free-Trade Agreements
FTAAs	Free Trade Areas of the Americas
GATT	General Agreement on Tariffs and Trade
GCP	Guidelines for Good Clinical Practice
GNP	Gross National Product
HGP	Human Genome Project
HUGO	Human Genome Organization
IBC	International Committee on Bioethics
ICCPR	International Covenant on Civil and Political Rights
ICESCR	International Covenant on Economic, Social and Cultural Rights
ICH	International Conference on Harmonisation
IFIs	International Financial Institutions
IMF	International Monetary Fund
Int J Hum Rights	International Journal of Human Rights
Int'l L.	International Law
IP	Intellectual Property
IPRs	Intellectual Property Rights
ITPGRFA	International Treaty on Plant Genetic Resources for Food and Agriculture
MDGs	Millennium Development Goals
MRC	Medical Research Council
R&D	Research & Development
SNPs	Single Nucleotide Polymorphisms
TCC	Transnational Capitalist Class
TNCs	Transnational Corporations
TRIPS	Trade Related Aspects of Intellectual Property

UDHR	Universal Declaration of Human Rights
UN	United Nations
UNESCO	United Nations Educational, Scientific and Cultural Organization
UNICEF	United Nations Children's Fund
WHO	World Health Organization
WIPO	World Intellectual Property Organization
WMA	World Medical Association
WTO	World Trade Organization

Foreword

The issue of access to medicines, be it in respect of HIV/AIDS, malaria, tuberculosis or any of a multitude of tropical diseases, is a constant refrain in international fora dealing with health, development and the global economy. At the 2009 G8 meeting in l'Aquila, Italy, world leaders promised, for example, to 'implement further efforts towards universal access to HIV/AIDS prevention, treatment and care and support by 2010'. Similar sentiments have been expressed at the G20, the World Health Organization, the World Trade Organization and at the World Intellectual Property Organization. Never before has there been such broad and sustained political support for the principle of global access to medicines. Yet, these same leaders have continually failed to deliver on their statements or to provide the aid they promised.

At the same time, particularly after the 2008–2009 economic crisis, critics point to the world's poor record on humanitarian assistance. Some, such as Dambisa Moyo, go so far as to say that current Western aid programmes have increased corruption and dependence, deepening rather than alleviating the plight of the world's poor. Beyond that, Western countries routinely make pledges of aid that they never fulfill. As with the access to medicines debate, there is an increasing disparity between what countries say they are doing and the effects of their actions.

Debates over access to medicines and humanitarian assistance are deeply interconnected and oppositional. The failure to deliver needed medicines is taken as evidence that intellectual property rules, and patents in particular, contradict the principles of human rights since they put profits ahead of people. On the other hand, the inability to build sustainable economies in so many developing countries is seen as a failure of human rights. According to this critique, only a more market-oriented system, complete with intellectual property rights that encourage private investment, has the capacity to build such economies.

The oppositional nature of the debate between intellectual property and human rights, particularly between the private sector and civil society, has had some positive outcomes. Drawing on human rights arguments, civil society has not only put the issue of access to medicines on the international agenda but has succeeded in having the World Trade Organization adopt measures to facilitate access despite the existence of patent rights. This in turn has spurred industry to establish drug donation programmes which further encouraged civil society to push for more substantive reforms of patents and its alternatives.

Despite these positive outcomes, the opposition between human rights and intellectual property – concretely, between civil society and industry – has left millions of people without needed medicines and has had only limited success in encouraging research on the diseases that overwhelmingly affect the world's poor. If one truly wishes to address both the economic and health conditions in developing countries, one would need to overcome this opposition.

This is, in short, the project of Louise Bernier in this book. Instead of accepting the assumption that intellectual property and human rights are incompatible, Bernier provides a common starting point for analysis of each. Drawing on a cosmopolitan theory of justice, Bernier shows how both intellectual property and human rights law share a common instrumental role in establishing a just society. While she finds faults with each, both are best understood, she argues, as different instantiations of a common vision of justice.

This understanding of intellectual property and of human rights – as instruments in attaining justice – provides for a common metric for the analysis of both. Bernier examines the extent to which each of intellectual property and human rights advances justice. The result is surprising to one who believes in the primacy of one over the other: that both advance and detract from justice equally. Only by combining both sets of laws can we truly advance the cause of access to medicines.

Moving us to the point of attempting to reconcile these disparate sets of laws is a critical first step. Bernier, at the end of this book, sketches out possible pathways to undertake that reconciliation. It is up to the rest of us to build on those pathways and to explore others.

Richard Gold

Preface

Genetics is one sector in which there has been tremendous evolution and progress over the last few decades. While it is believed that genetics could offer tremendous opportunities for global health improvement, there is also a fear that existing global health inequalities will be amplified by the evolution of genetics.

It thus appears necessary to analyse the way current assumptions define what is just and acceptable with regard to global access and distribution of resources in this field. Indeed, given the importance of genetics to human health globally, this book will evaluate two principal legal regimes – intellectual property and international human rights – to determine to what extent they further the goal of distributing the benefits of these technologies equitably and globally. This evaluation is vital to ensure that legal regimes assist in ensuring that this promising field develops in a way that improves global health without leaving the most vulnerable outside of the process. This book will undertake this complex task by employing and building upon cosmopolitan liberal theories developed over the few last decades as an extension of the work of Rawls and Daniels.

A theoretical framework to justify engaging in a global and more equitable redistribution of benefits produced by genetics is required. Ultimately, my analysis will produce strong normative benchmarks based on justice considerations for engaging in a global and more equitable redistribution of the benefits likely to emerge from genetic science. Universal consideration of all human beings, importance of health needs, normal functioning and equality of opportunities are some of the notions that will be analysed to construct this framework. I will then attempt to determine how and if this theory of distribution translates into positive law and to identify and analyse the main obstacles to legal compliance with global distributive justice. I will assess two main international normative systems: intellectual property law and human rights law to determine if their underlying philosophy, structure, and functioning take account of the principles highlighted in my theoretical framework and how underlying politics and economics matter.

This will set out a basis for further discussion on how we could work around some of the major obstacles identified throughout my analysis. It will also help us move from the vague and often symbolic ideal of benefit sharing actually prevailing toward the establishment of a real, enforceable concept of global benefit sharing in health that would position genetics at the rank of an essential tool for achieving global health.

Introduction

Of all forms of inequality, injustice in health care is the most shocking and inhumane.

Martin Luther King, Jr.[1]

THE GLOBAL HEALTH CHALLENGE

Today's world is characterised by a disturbing reality: on one hand, there is remarkable and ongoing technological progress in various spheres of activity while, on the other, the substantial gap existing between the world's rich and its poor constantly deepens. Indeed, despite our impressive state of knowledge, innovation, and development, at least 1.1 billion individuals continue to fight for their daily survival, more than 2.7 billion others live in acute poverty, on less than $2 a day and more than 200 million children are prevented from achieving their full development potential.[2] Such extreme poverty engenders terrible consequences, such as widespread infant mortality and adult premature deaths, severe malnutrition, and lack of access to basic necessities such as drinkable water, basic sanitation, shelter, and health care.[3] All UN Member

1 Quoted in L. Sheremeta and B.M. Knoppers, 'Beyond the Rhetoric: Population Genetics and Benefit-Sharing' (2003) 11 *Health Law Journal* 89.

2 Early Child Development Knowledge Network of the Commission on Social Determinants of Health, *Total environment assessment model for early child development – Evidence report*, 2007, Geneva, World Health Organization, online http://www.who.int/social_determinants/resources/ecd_kn_evidence_report_2007.pdf (accessed 20 May 2009) at p. 86; P.L. Engle et al. 'Strategies to Avoid the Loss of Developmental Potential in More than 200 million Children in the Developing World' (2007) 369 *Lancet* 229; Commission Social Determinants of Health (CSDH), *Closing the Gap in a Generation: Health Equity through Action on the Social Determinants of Health*, 2008, Geneva, World Health Organization, online: http://whqlibdoc.who.int/publications/2008/9789241563703_eng.pdf (accessed May 2009).

3 Pogge reports that, every day, 50 000 people die from poverty-related causes such as starvation, tuberculosis, malaria, and diarrhoea, and that even if those conditions affect 20% of the world population, they receive only 0.3% of all research funds: T.W. Pogge, 'Human Rights and Global Health: A Research Program' (2005) 36:1/2 *Metaphilosophy* 182, at 197; T.W. Pogge, 'Recognized and Violated by International

States acknowledged the seriousness of the situation in September 2000 when they adopted the *United Nations Millennium Declaration* that included commitments to encourage development, decrease poverty, and improve people's living conditions and health by 2015.[4] However, in a report published in 2008, the UN confirmed that more than halfway towards the target date of 2015, 'we are not on track to fulfil our commitments'.[5]

One of the greatest enduring problems in the world is the major disease burden affecting a large portion of the world population. The most important and serious health variations between individuals are not, for the most part, associated with biological and genetic determinants but with patterns of resources distribution.[6] As powerfully put by the Commission on the Social Determinants of Health (CSDH) in its final report: 'health and illness follow a social gradient: the lower the socioeconomic position, the worse the health'.[7] Indeed, health, diseases and the burden of disease are directly influenced by many factors including geographic situation, governments' commitment towards health improvement, scientific capacity, public infrastructures, health research investment and availability of financial, material, and human resources, which differ greatly across socio-economic groups and countries.[8]

Law: The Human Rights of the Global Poor' (2005) 18:4 *Leiden Journal of International Law* 717; see also UN Development Programme, *Human Development Report 2004* (New York: Oxford University Press, 2004), at 129–130; S. Chen and M. Ravallion, 'How Have the World's Poorest Fared since the Early 1980s?' (2004) *World Bank Research Observer* 153; UNICEF, *The State of the World's Children 2005* (New York: UNICEF, 2005).

4 UN General Assembly Resolution, *United Nations Millennium Declaration*, September 2000, A/RES/55/2, online on the UN website, http://www.un.org/millennium/declaration/ares552e.pdf (accessed 20 May 2009).

5 United Nations, *The UN Millenium Developement Goals Report*, New York, 2008, online http://www.un.org/millenniumgoals/pdf/The%20Millennium%20Development%20Goals%20Report%202008.pdf (accessed 20 May 2009), p. 3.

6 A.K. Acharya, 'Toward Establishing a Universal Basic Health Norm' (2004) 18:3 *Ethics and International Affairs* 65.

7 See the executive summary of CSDH, *supra* note 2.

8 Indeed, this reality has been called the 10/90 gap, where 90% of all health research gets dedicated to the most affluent 10% of the world. For more on this, refer to E. Dowdeswell, A.S. Daar and P.A. Singer, 'Bridging the Genomics Divide' (2003) 9 *Global Governance* 1; see also D.H. Peters et al., 'Reducing the Impact of Poverty on Health and Human Development: Scientific Approaches', Annals of the New York Academy of Sciences (2008) Vol. 1136, 161; J. Coloma and E. Harris, 'Sustainable Transfer of Biotechnology to Developing Countries – Fighting Poverty by Bringing Scientific Tools to Developing-Country Partners', Annals of the New York Academy of Sciences (2008) Vol. 1136, 358; R.A. Malkin, 'Design of Health Care Technologies for the Developing World' (2007) 9 *Annual Review of Biomedical Engineering* 567; World Health Organization, *World Health Report 1999: Making a Difference in People's Lives: Achievements and Challenges*, Geneva, 1999; T. Evans et al., *Challenging*

The impact of economic inequality on people's health can be observed both within and between countries and depends on various factors including governmental resources and priorities and the availability and affordability of universal or private health insurance coverage. Even if we do not question the importance and seriousness of health inequities arising at the national level, this book will focus on the growing global health divide between populations from the nations of the North and the South[9] and on the possibility that this divide will be aggravated by the introduction of genetic technologies aimed at health improvement.

WHY GENETICS?

Genetics is only one among many spheres in which we are likely to continue being confronted with gross inequalities in health but it is a sector in which there has been tremendous evolution and progress over the last few decades. While it is believed that genetics could offer tremendous opportunities for global health improvement and play an important role in meeting the UN *Millennium Development Goals* (MDGs), there is also a fear that existing global health inequalities will be amplified by the evolution of genetics. Such a divide already exists in relation to numerous vital health-related determinants such as nutrition, water, shelter and labour.[10] In this sense, this book does not provide a complete picture of global health inequalities but focuses

---

Inequities in Health: From Ethics to Action (New York: Oxford University Press, 2000).

Concerning this reality in the research in genetics see: J. Enriquez, R. Martinez and J. West, 'The New World Order', (2008) 20 Flypmedia, online: flypmedia.com/issues/20/#6/3 (accessed 21 May 2009).

[9] I acknowledge that different developing countries are at varying stages of development and that some of the issues they face might be different depending on the country. The *Economic and Social Council of the United Nations* has identified some countries with the label 'least developed countries' based on different criteria such as low income, human resources weakness, and economic vulnerability. Other countries not part of this latter category are nevertheless characterised as developing countries given their level of development and a plethora of different socio-economic, demographic, and political factors. For the purposes of this book, I adopt an inclusive notion of developing countries, taking different degrees of poverty, lack of resources, and health access problems as my general benchmark. Throughout this book, the expression 'developing nations' or 'developing world' should therefore be read to include a range of countries, mostly southern, in need of more health resources, including the least developed ones as well as those who are at a medium level of development.

[10] I therefore follow a multi-causal conception of health under which health is influenced by biological, social, economic, psychological, environmental, and genetic factors.

on one important sphere of activity and on how it can be harnessed and developed to improve global health.

In less than 15 years, our understanding of genetics has evolved considerably in various areas such as agriculture, biodiversity, traditional knowledge, biomedical research, and medical applications.[11] As clearly put by Hinojosa, '[f]ew developments in science have had the impact on society, institutions, laws, and health care that genetics is having and, undoubtedly, will continue to have.'[12] In the field of human genetics, we went from knowing very little of the particulars of biological genetics to a situation where a tremendous amount of information about the structure of individual genes is discovered daily.[13] Six years ago, *Nature* and *Science* published two series of articles highlighting the great potential of genetics and the need to do much more in terms of discovery and analysis of gene and protein functions, interactions, and their role in diseases, conditions, and reactions.[14] More recently, the

---

11 Indeed, some genetic resources are already used for agriculture, medicine, and industrial development both in developed and developing countries. However, since this book focuses on the global medical promises of human genetics and the distribution arising from the development of genetics at this level of activity, it is beyond its scope to address, in detail, the fields of agriculture, plant genetics, and traditional knowledge. I am, however, aware that many issues addressed here also find application in other genetics-related sectors of crucial importance for human health.

12 J.P. Hinojosa, 'The Human Genome, Property of All: Opportunities Under the ALRC Inquiry into Gene Patenting and Human Health' (2004) 26 *Sydney Law Review* 447, at 448.

13 The *Human Genome Project* (HGP) was launched in the early 1990s to 'determine the complete sequence of the three billion DNA (molecules encoding genetic information) subunits (bases), identify all human genes, and make them accessible for further biological study'. It marked the beginning of a new age in science. In June 2000, Francis Collins from the HGP and Craig Venter from *Celera Genomics* simultaneously announced the completion of a first working draft of the human genome sequence, and in February 2001, both groups published their initial draft map independently. This sequencing disclosed much information on the number of human genes (about 30 000 instead of the first estimated 80 000–100 000) and their composition, and helped to identify many other interesting biological mutations, including more than two million genetic variations (*single nucleotide polymorphisms: SNPs*). For a few references on the development of genetics see: L. Peltonen and V.A. Mckusick, 'Dissecting Human Disease in the Postgenomic Era' (16 February 2001) 291:5507 *Science* 1224; Human Genome Project Information, *Frequently Asked Questions*, on line on the HGP website http://www.ornl.gov/sci/techresources/Human_Genome/faq/faqs1.shtml (accessed 3 March 2009); J.C. Venter et al., 'The Sequence of the Human Genome' (16 February 2001) 291:5507 *Science* 1304; International Human Genome Sequencing Consortium, 'Initial Sequencing and Analysis of the Human Genome' (15 February 2001) 409 *Nature* 860.

14 F.S. Collins, M. Morgan and A. Patrinos 'The Human Genome Project: Lessons from Large-Scale Biology' (11 April 2003) 300: 5617 *Science* 286; M.E.

science of genetics has been taken to a whole new level when some of the world's most famous geneticists announced that they were using genetics to create artificial DNA sequences from scratch and that their next step was to implant this manufactured DNA into a cell.[15] If successful and adequately framed and supervised, these initiatives might provide the basis and the tools necessary for developing promising new therapeutic approaches and techniques with the potential to prevent, screen, and cure very serious diseases. Scientists thus face numerous and exciting challenges in this area.[16]

Genetic factors play some role in almost all human diseases. These factors either confer susceptibility, resistance or influence individuals' interactions with their environment. For many years, genetics has been critical in revealing the cause of certain monogenic diseases.[17] However, things become more difficult when we try to establish connections between individual genotypes and complex diseases involving many genes and environmental factors such as hypertension, cancer, or schizophrenia.[18] This is where genomics, the 'study of genes and their function',[19] becomes extremely useful and important. It allows broad analysis of numerous genes simultaneously to obtain a better idea of how they interact with one another and become expressed in specific cell types.[20] Attention is increasingly focused on using a combination of genetics, genomics and cutting edge software tools to develop 'sophisticated

---

Frazier et al., 'Realizing the Potential of the Genome Revolution: The Genomes to Life Program' (11 April 2003) 300: 5617 *Science* 290; F.S. Collins et al., 'A Vision for the Future of Genomics Research' (24 April 2003) 422 *Nature* 835.

15 D.G. Gibson et al., 'Complete Chemical Synthesis, Assembly, and Cloning of a Mycoplasma genitalium Genome' (29 February 2008) 319:5867 *Science* 1215; Jürgen Pleiss, 'The Promise of Synthetic Biology' (2006) 73 *Appl. Microbiol. Biotechnol.* 735; P. Ball, 'Synthetic Biology: Starting from Scratch' (2004) 431 *Nature* 624.

16 B. Albert and A. Klug, 'The Human Genome Itself Must be Freely Available to all Humankind' (23 March 2000) 404 *Nature* 325; Business Week, 'On The Brink Of Artificial Life' (25 June 2007), online http://www.businessweek.com/magazine/content/07_26/b4040047.htm (accessed 2 April 2009).

17 A.D. Roses, 'Pharmacogenetics and the Practice of Medicine' (15 June 2000) 405 *Nature* 857.

18 N.A. Holtzman and T.M. Marteau, 'Will Genetics Revolutionize Medicine' (13 July 2000) 343:2 *New England Journal of Medicine* 141.

19 Human Genome Project Information, *Genome Glossary*, on line on the HGP website, http://www.ornl.gov/sci/techresources/Human_Genome/glossary/glossary_g.shtml (accessed 3 March 2009).

20 N. Bhardwaj and H. Lu, 'Correlation between gene expression profiles and protein–protein interactions within and across génomes' (2005) 21:11 *Bioinformatics* 2730; M.M. Hopkins et al., 'Putting Pharmacogenetics into Practice' (2006) 24 *Nature Biotechnol.* 403; M. Mowzoon, 'Access Versus Incentive: Balancing Policies in Genetic Patents' (2003) 35 *Ariz. St. L.J.* 1077.

microarray technologies' that could be used to screen for complex diseases, achieve better cellular and molecular understanding of those conditions and develop new therapeutic tools.[21] In this book, I understand the science of genetics in a broad sense, as an entire field of activity that includes the interaction between functional genomics, new computational analytical methods, proteomics, traditional genetic testing and screening techniques, and the understanding and conversion of the data emerging from this research into practical and useful applications to improve global health.

Also, for the purposes of this book, I will concentrate on those aspects of health that are common to all individuals. This universal perspective is therefore not adjustable to personal circumstances and does not vary with each individual's perception and preferences. I will focus on universal and objective human health needs, leaving questions pertaining to the enhancement of otherwise normal traits to others.[22] This definition of health thus relates to *normal functioning* as opposed to a perfectionist conception of healthy human beings, where normal functioning is the objective capacity of individuals to take advantage of a reasonable range of opportunities. In this sense, genetics represents one tool among many to satisfy essential medical needs and to help bring all individuals to a universal minimal health level under which they can expect a 'decent' life.

Nevertheless, the science of genetics has some limits as most diseases are caused by a variety of factors and by complex interactions between genes and the environment.[23] In response to the growing enthusiasm for the genetic revo-

---

[21] Up to now, numerous polymorphisms influencing how one responds to and metabolises certain drugs have been identified with novel sequencing and bioinformatics methods. Also, new vaccines arising from pathogen DNA are being developed and progress has been made in understanding cancer mechanisms with research in genetics. Indeed, the most important mutations have been identified in a family of 'cellular oncogenes', and the next step is now to find the specific genes associated with the more common cancers with wide genomics investigations. For more information on those technologies, refer to: T. Joos and P. Kroeger, 'New Frontiers in Microarray Technology Development' (2008) 19:1 *Current Opinion in Biotechnology* 1; WHO, *Genetics, Genomics and the Patenting of DNA: Review of Potential Implications for Health in Developing Countries*, Geneva, 2005, online on the WHO website, http://www.who.int/genomics/FullReport.pdf (accessed 23 February 2009).

[22] For more on the role of genetics in enhancement, I refer the reader to: F. Allhoff, 'Germ-Line Genetic Enhancement and Rawlsian Primary Goods' (May 2008) 18:1 *Journal of Evolution and Technology* 10; N. Bostrom, 'Human Genetic Enhancements: A Transhumanist Perspective' (2003) 37:4 *Journal of Value Inquiry* 493; L.B. Andrews, *Future Perfect* (New York: Columbia University Press, 2001); A. Buchanan et al., *From Chance to Choice: Genetics and Justice* (Cambridge: Cambridge University Press, 2000); T. Peters, *Playing God: Genetic Determinism and Human Freedom* (New York: Routledge, 1997).

[23] R.G. Ramos and K. Olden, 'Gene-Environment Interactions in the Development of Complex Disease Phenotypes' (March 2008) 5:1 *Int. J. Environ Res*

lution, some observers suggest that the excitement surrounding genetics is, at times, over-stated and that excess of optimism should be moderated.[24] There are still many technological and statistical obstacles to overcome in linking phenotypes to genetic markers, and some believe that progress will take significantly more time.[25] Further, clinical applications of genetic knowledge sometimes remain limited, even when much information is available.[26]

One must therefore come to this subject by realising that people working in genetics are still in the early phase of understanding the complexity of gene interactions. Nevertheless, significant progress has been made in a very short time. There are many signs that valuable technological development will continue to take place and that genetics will continue to have an important preventive and therapeutic role to play in health care, medical practice and public health.[27] Therefore, although I realise that it might take many more years and significant investment to get to a point where genetics can fully deliver on its promises, this book focuses on the progress already made, starting from the premise that it is only a matter of time before technical challenges are overcome and genetic research can create greater benefits for the delivery of health care on a global scale.[28]

Public Health 4; J. Alper, 'Genetic Complexity in Human Disease and Behavior' in J. Alper et al. (eds), *The Double-Edged Helix: Social Implications of Genetics in a Diverse World* (Baltimore: Johns Hopkins University Press, 2002) 17; E. T Juengst 'FACE Facts: Why Human Genetics Will Always Provoke Bioethics' (Summer 2004) 32:2 *Journal of Law, Medicine and Ethics* 26.

24 T.M. Bubela and T. Caulfield, 'Media Representations of Genetic Research' in E.F. Einsiedel and F. Timmermans (eds), *Crossing Over. Genomics in the Public Arena* (Calgary: University of Calgary Press, 2005); T.M. Bubela and T. Caulfield, 'Does the Print Media Hype Genetic Research?: A Comparison of Newspaper Stories and Peer Reviewed Research Papers' (2004) 170:9 *Canadian Medical Association Journal* 1399; S. Jones, *Genetics in Medicine: Real Promises, Unreal Expectations: One Scientist's Advice to Policymakers in the United Kingdom and the United States* (London: Milbank Memorial Fund, 2000); L.B. Andrews, 'Past as Prologue: Sobering Thoughts on Genetic Enthusiasm' (1997) 27 *Seton Hall L. Rev.* 893.

25 J. Altmuller et al., 'Genomewide Scans of Complex Human Diseases: True Linkage is Hard to Find' (2001) 69 *American Journal of Human Genetics* 936; D.S. Roos, 'Bioinformatics – Trying to Swim in a Sea of Data' (2001) 291 *Science* 1260.

26 B.R. Bloom and D.D. Trach, 'Genetics and Developing Countries' (April 28, 2001) 322 *BMJ* 1006; N. Holtzman, 'Will Genetics Revolutionize Medicine?' (2000) 343 *New England Journal of Medicine* 141.

27 G. Wang and C. Watts, 'The Role of Genetics in the Provision of Essential Public Health Services' (2007) 97:4 *Am. J. Public Health* 620; F.S. Collins et al., *supra* note 14; A.E. Guttmacher and F.S. Collins, 'Genomic Medicine – a Primer' (2002) 347:19 *New England Journal of Medicine* 1512; H. Varmus, 'Getting Ready for Gene-Based Medicine' (2002) 247 *N. Eng. J. Med.* 1526.

28 WHO, The Advisory Committee on Health Research of the World Health Organization, *Genomics and World Health*, Geneva, 2002, online on the WHO website, http://www3.who.int/whosis/genomics/pdf/genomics_report.pdf (accessed 4 March 2009).

INFLUENCE OF GENETICS ON GLOBAL HEALTH

Even if genetic innovation has mainly occurred in the developed world because it requires high capital investment, cutting edge technology, and well-equipped infrastructure, this does not mean that genetics does not have the potential to help the less affluent.[29] Indeed, studies reveal that human genetics offers a number of targeted possibilities for improving health in the developing world such as through the use of molecular diagnosis for better management and screening of infectious, non-infectious and parasitic diseases, and through new drug and vaccine development.[30]

Moreover, in the clinical setting, genetic testing can be used to address the specific health needs of the developing world's populations. Numerous existing genetic services could be beneficial in the developing world where chronic non-communicable diseases 'are reaching epidemic proportions [and] [t]he number of deaths from these diseases is double the number of deaths that result from a combination of infectious diseases (including HIV/AIDS, tuberculosis and malaria), maternal and perinatal conditions, and nutritional defi-

---

[29] Address by Dr Gro Harlem Brundtland, Director-General to the Fifty-fourth World Health Assembly, *Bridging the Health Divide: The Way Forward*, May 14, 2001, online on the WHO website, www.who.int/director-general/speeches/2001/english/20010514_wha54.html (accessed 26 February 2009).

[30] B. Séguin et al., 'Human Genomic Variation Initiatives in Emerging Economies and Developing Countries' (October 2008) *Nature Reviews, Genetics – Genomic Medicine in Developing Countries*, online www.nature.com/nrg/supplements/genomicmedicine (accessed 26 April 2009); W.H. Long et al., 'A Universal Microarray for Detection of SARS Coronavirus' (2004) 121 *J. Virol. Methods* 57; R.K. Hui et al., 'Reverse Transcriptase PCR Diagnostic Assay for the Coronavirus Associated with Severe Acute Respiratory Syndrome' (2004) 42 *J. Clin. Microbiol* 1994; R. Kukreti et al., 'Beta(2)-adrenergic Receptor Polymorphisms and Response to Salbutamol among Indian Asthmatics' *Pharmacogenomics* 6, 399–410 (2005). University of Toronto Joint Centre for Bioethics, *Top 10 Biotechnologies for Improving Health in Developing Countries*, Toronto, 2002; WHO, The Advisory Committee on Health Research of the World Health Organization, *Genomics and World Health*, Geneva, 2002, online on the WHO website, http://www3.who.int/whosis/genomics/pdf/genomics_report.pdf (accessed 4 March 2009), on concrete possibilities for the development of new vaccines, diagnostics, and therapeutic tools, refer especially to section 2 and 3; A.S. Daar et al., 'Top Ten Biotechnologies For Improving Health In Developing Countries' (October 2002) 32 *Nature Genetics* 269. This study identifies the ten most promising biotechnologies for improving health in developing countries in the next decade. Out of ten, six are directly or indirectly related to genetics; see also P.A. Singer and A.S. Daar, 'Harnessing Genomics and Biotechnology to Improve Global Health Equity' (5 October 2001) 294 *Science* 87; A. Buchanan, *Justice, Legitimacy and Self-Determination: Moral Foundations for International Law* (Oxford: Oxford University Press, 2004) chapter 4, p. 191.

ciencies'.[31] Indeed, a number of severe and life-threatening non-communicable diseases with a strong genetic component could likely be prevented, recognised, diagnosed, and treated in the future if safe, targeted and efficient drug development and genetic-predisposition testing were made available.[32] Such preventive strategies could be especially beneficial in addressing situations where neither individuals nor governments are able to pay for costly and lengthy treatments.[33]

Despite many economic, structural and political obstacles, genetic research is nevertheless being conducted in some countries of the developing world. Indeed, recent case studies highlight genotyping initiatives undertaken in

---

[31] A.S. Daar et al., 'Grand Challenges in Chronic Non-Communicable Diseases' (2007) 450 *Nature* 494; A. Boutayeb, 'The Double Burden of Communicable and Non-Communicable Diseases in Developing Countries' (2006) *Trans. R. Soc. Trop. Med Hyg* 191; O. Adeyi, O. Smith and S. Robles, 'Public Policy and the Challenge of Chronic Noncommunicable Diseases', World Bank, Washington DC, 2007; D.G. Richards, *Intellectual Property Rights and Global Capitalism, The Political Economy of the TRIPS Agreement* (London: M.E. Sharpe, 2004) chapter 6, p. 141.

[32] For example, the Indian Genome Variation Database Consortium recently linked a genetic polymorphism (haplotype at the ADRB2 locus) with asthma patients' reactions to salbutamol. For more on this research initiative, see P. Bhatnagar, R. Guleria and R. Kukreti, 'Variable Therapeutic Response in Asthma: a Genetic Perspective' (2006) 3 *Personalized Medicine* 61; R. Kukreti et al., 'Beta(2)-adrenergic Receptor Polymorphisms and Response to Salbutamol among Indian Asthmatics' *Pharmacogenomics* 6, 399–410 (2005). Moreover, the application of diagnostic measures that use DNA analysis to identify genetic carriers or diseases could be very helpful to screen for red blood cell disorders like thalassaemia and sickle cell disorders (very common in developing countries) and inform carriers of the risk to their health and their offspring's health. Another example is prenatal diagnosis, which has been found to be useful in the identification of sickle cell anaemia, a very serious condition associated with a high level of mortality and morbidity. Access to this measure would be of particular interest to allow preventive action in West Africa where almost 25% of the population are sickle cell carriers. Early molecular diagnosis and neonatal screening for Cystic Fibrosis (CF) is a third example of preventive genetic medicine that could be of great value for the developing world. In Brazil, where there is a high incidence of CF, it is not rare that persons afflicted with the disease die undiagnosed. Therefore, basic genetic testing could be extremely useful in Brazil and other countries with similar rates of CF, especially for lower-income families. However, for now, those existing diagnostic tools do not reach the bulk of the world's population who need it the most, particularly those living in remote rural areas of developing countries. A. Alwan and B. Modell, *Community Control of Genetic and Congenital Disorders*, EMRO Technical Publications, Series 24, World Health Organization, Regional Office for the Eastern Mediterranean, Cairo, 1997; WHO, *Guidelines for Control of Haemoglobin Disorders*, Geneva, 1994, WHO/HDP/HB/GL/94.1; WHO, *supra* note 30, at 81–84; C. Streit et al., 'CFTR gene: Molecular Analysis in Patients from South Brazil' (2003) 78 *Molecular Genetics and Metabolism* 259; M. Petrou and B. Modell, 'Prenatal Screening for Haemoglobin Disorders' (1995) 15 *Prenat. Diagn* 1275.

[33] A. Alwan and B. Modell, op cit at 62.

Mexico, India and Thailand to unveil patterns of genetic variations between individuals and to better understand disease susceptibility in populations.[34] This literature however reveals that despite strong political will and increased capacity building in some countries, limited funding, normative obstacles and poor infrastructures prevent most developing nations from pushing genetics further and translating findings into concrete actions to benefit their populations.

Consequently, given the fact that genetics is a very promising field for helping to improve global health – and considering that it has not, up to now, been primarily developed with this focus – I consider it essential to discuss how the field's emerging benefits (knowledge, expertise, research tools, products and services, and profits) should be better produced and distributed on the global scene in the future.[35] There are a plethora of normative, socio-economic, structural and political obstacles to more equitable distribution of health-related benefits and I will address some of these in the course of this book. However, before I go any futher, as my primary focus is to argue for more equity in access to genetics' emerging benefits, it is important to say a few words on the popular concept of *benefit sharing* and see if, as it stands now, it provides an adequate response to the widening global health divide.

BENEFIT SHARING

The expression *benefit sharing* is used broadly in relation to biodiversity, traditional knowledge, and human genetic research to indicate that some of the benefits (economic or social) arising from these fields should be shared with those from whom the goods or knowledge originated.[36] For example, the

---

[34] B. Séguin et al., 'Genomic Medicine and Developing Countries: Creating a Room of their Own' (June 2008) 9 *Nature Review Genetics* 487.

[35] B.R. Bloom and D.D. Trach, *supra* note 26.

[36] For example, one of the main objectives of the *Convention on Biological Diversity* (CBD) is the fair and equitable sharing of benefits arising from the knowledge deriving from biological diversity. The *Bonn Guidelines* on access to genetic resources and fair and equitable sharing adopted in April 2002 provide a strategy for the access and benefit sharing process and the Food and Agriculture Organization (FAO) Conference approved the *International Treaty on Plant Genetic Resources for Food and Agriculture* (ITPGRFA), which entered into force in June 2004 and by which contracting parties will provide easier access to identified genetic resources through a unique *Multilateral Material Transfer Agreement*. For reference to those legal documents, see: *Convention on Biological Diversity*, Rio de Janeiro, 5 June 1992, online: http://www.biodiv.org/doc/legal/cbd-en.pdf (accessed 4 May 2009); Sixth Meeting of the Conference of the Parties to the Convention on Biological Diversity, Decision

Human Genome Organisation (HUGO) issued a statement in 2000 calling for the sharing of certain benefits arising from the commercialisation of genetic inventions with the populations or communities who first donated their samples.[37] Generally, benefit-sharing obligations arise from two different situations: as a consequence of specific transactions with research participants and contributors, or as a result of a norm that the good in question ought to be used for the general benefit of all humanity.[38]

The first and most popular application of this obligation involves sharing the benefits of research with the contributors of genetic resources based on an assumption that these people are entitled to share in the benefits in a concern for compensatory justice.[39] The actual justification put forward for this obligation varies depending on whether one is discussing human biological samples, biodiversity and plants, or traditional knowledge. I will not address each of these in detail as it is beyond the scope of this work. However, to get a better understanding of the compensatory argument, it is useful to say a few words on the rationale for sharing benefits arising from the use of biological tissues when they are provided by individuals and populations for human genetic research.[40]

VI/24, *Access and Benefit-Sharing as related to genetic resources*, April 2002, the Hague, online: http://www.biodiv.org/decisions/default.asp?m=cop-06andd=24 (accessed 4 May 2009); FAO, *International Treaty on Plant Genetic Resources for Food and Agriculture*, November 2001, online: ftp://ftp.fao.org/ag/cgrfa/it/ ITPGRe. pdf; For an interesting discussion on the concept of benefit sharing as applied to human genetics, see: K. Simm, 'Benefit-Sharing Regarding the Meaning and Limits of the Concept in Human Genetic Research' (2005) 1:2 *Genomics, Society and Policy* 29.

37 HUGO, *Statement on Benefit Sharing*, Vancouver, 2000, online: http://www.gene.ucl.ac.uk/hugo/benefit.html (accessed 4 May 2009).

38 This dual aspect is clearly highlighted and explained in this UNESCO report: UNESCO International Committee on Bioethics, *Report of the IBC on Ethics, Intellectual Property and Genomics*, 10 January 2002, SHS-503/01/CIB-8/2 Rev.

39 For example, HUGO, *Statement on Benefit Sharing*, *supra* note 37, E. Justice, 1) Compensatory justice: meaning that the individual, group, or community, should receive recompense in return for contribution; Convention on Biological Diversity, *supra* note 36, art. 15. For more on this application see Participant in the 2001 Conference on Ethical Aspects of Research in Developing Countries, 'Fair Benefits for Research in Developing Countries' (13 December 2002) 298 *Science* 2133; UN Commission on Human Rights, High Commissioner, 'The impact of the Agreement on Trade-Related Aspects of Intellectual Property Rights on Human Rights', Geneva, 27 June 2001.

40 For more on the justification for providing compensation (or not) when using biodiversity and plants' genetic resources and traditional knowledge, I refer the reader to: L. Mansur, 'Gene Discovery, Ownership and Access for Developing Countries in the Era of Molecular Genetics' (2002) 5:1, *Electronic Journal of Biotechnology*, online: http://www.ejbiotechnology.info/content/vol5/issue1/issues/05/ (accessed 16 May

The provision of biological material is regulated by the broad principle of non-commercialisation of the human body and its components[41] illustrated by the prevalence of a 'consent model' (under which individuals are entitled to give away or abandon bodily materials), in contrast to a 'property model' (under which individuals are entitled to sell their tissues).[42] There is also a debate as to whether the current system should be revisited to allow property

2009); UK Commission on Intellectual Property Rights, *Integrating Intellectual Property Rights and Development Policy*, London, September 2002, at 84; Indian Government, WTO, *Protection of Biodiversity and Traditional Knowledge. The Indian Experience*, 14 July 2000, WT/CTE/W/156; *International Convention of the Protection of New Varieties of Plants* (UPOV), Paris, 1961 and revised in Geneva in 1972, 1978 and 1991, online on the UPOV website, http://www.upov.int/en/publications/conventions/index.html (accessed 8 March 2009); World Intellectual Property Organization (WIPO), Intergovernmental Committee on Intellectual Property and Genetic Resources, Traditional Knowledge and Folklore, *Operational Terms and Definitions*, Geneva, 20 May 2002, WIPO/GRTKF/IC/3/9; WIPO, *The Protection of Traditional Cultural Expressions/Expressions of Folklore: Revised Objectives and Principles*, Geneva, January 2006, WIPO/GRTKF/IC/9/4, online on the WIPO website:http://www.wipo.int/edocs/mdocs/tk/en/wipo_grtkf_ic_9/wipo_grtkf_ic_9_4.pdf (accessed 8 March 2009); C. Correa, *Traditional Knowledge and Intellectual Property*, The Quaker United Nations Office (QUNO), Geneva, November 2001, online on the website of netamericas: http://www.netamericas.net/Researchpapers/ Documents/Ccorrea/Ccorrea2.pdf 9 (accessed 8 March 2009); WHO, *Report of the Inter-Regional Workshop on Intellectual Property Rights in the Context of Traditional Medicine*, Bangkok, 6–8 December 2000 (WHO/EDM/TRM/2001.1); E.R. Gold and D. Castle, *Traditional Knowledge and Benefit Sharing: From Compensation to Transation*, paper presented at ICABR 8th International Conference on Agricultural Biotech: International Trade and Domestic Production, Ravello, Italy, July 2004; UN Commission on Human Rights, High Commissioner, 'The Impact of the Agreement on Trade-Related Aspects of Intellectual Property Rights on Human Rights', Geneva, June 27, 2001, at para 41.

[41] For example, see UNESCO Universal Declaration on the Human Genome and Human Rights, Paris, 1997, art. 4.

[42] For example, see the case *Moore v Regents of the University of California* 794 P 2d 479 Cal SC 1990, where a physician obtained a commercially valuable patent over a patient's cell and in which the California Supreme Court found that the patient did not have property interest in his cells even if they contained his DNA. The Court instead awarded compensation for breach of fiduciary duty and lack of informed consent; for more on the 'consent' model vs. the 'property' model, refer to European Group on Ethics in Science and New Technologies (EGE), Opinion No. 16, *Ethical Aspects of Patenting Inventions Involving Human Stem Cells*, Brussels, 7 May 2002, sect.1.20, at 12, online on the website of the European Parliament, http://europa.eu.int/comm/european_group_ethics/avis3_en.htm (accessed 25 April 2009); Medical Research Council, *Human Tissue and Biological Samples for Use in Research – Operational and Ethical Guidelines*, London: MRC Ethics Series, April 2001, s. 2.2, online on the MRC website, http://www.mrc.ac.uk/pdf-tissue_guide_fin.pdf (accessed 9 March 2009).

claims in some human body material.[43] However, this book proceeds on the basis that, as argued by Gold, it is inappropriate to apply property discourse to human bodily materials since this discourse cannot appropriately deal with goods that are valuable chiefly for *non-economic* reasons. There are numerous different values that relate to health and human biological material (dignity, community, spirituality and so on), many of which cannot and should not be evaluated by the market.[44] This leads me to argue that we should not consider participants' biological tissues as commercial property, appropriately valuable, for which compensation should be awarded.[45] Therefore, even if sharing benefits with human genetic-resource contributors may seem intuitively right and equitable for many, it appears to lack a real normative basis with regards to compensatory justice and property law.

On a more practical level, a compensatory system unequally rewards contributions to individuals and communities. Some populations and individuals, because of their geographical situations, special environment, employment, or genetic makeup will be more 'interesting' than others for the purpose of specific research on genetic diseases, variations, and polymorphisms. In these circumstances, compensatory benefit-sharing systems can be viewed as a kind of lottery where the luckiest individuals and populations participate and win a portion of the benefits and others, also in great need but without similar resources, are left out of the process completely.

There are ongoing discussions about proposals to impose obligations of benefit sharing on scientists toward specific resource contributors based on compensatory justice. This topic is fascinating, but given that the purpose of this book is not to provide a deep analysis and critique of compensatory benefit sharing mechanisms, and that I am not convinced that this is the most equitable and justified way to think about sharing genetics' benefits, I leave it

---

43 E.R. Gold, *Body Parts: From Property Rights to Human Biological Materials* (Washington D.C.: Georgetown University Press, 1996); *Moore v Regents of the University of California* 794 P 2d 479 Cal SC 1990; K. Mason and G. Laurie, 'Consent or Property? Dealing with the Body and its Parts in the Shadow of Bristol and Alder Hey' (September 2001) 64:5 *The Modern Law Review* 725; W. Boulier, 'Sperm, Spleens and Other Valuables: The Need to Recognize Property Rights in Human Body Parts' (1995) 23 *Hofstra L. Rev* 705.

44 Refer to Gold's book for a detailed analysis of those issues: E.R. Gold, *Body Parts: From Property Rights to Human Biological Materials*, op cit at chapters 7, 8 and 9; see also Nuffield Council on Bioethics, *Human Tissue: Ethical and Legal Issues* (London: Nuffield Council on Bioethics, April 1995), ss. 9.14 and 13.25, online on the Nuffield Council website, http://www.nuffieldbioethics.org/fileLibrary/pdf/human_tissue.pdf (accessed 9 March 2009).

45 E.R. Gold and T.A. Caulfield, *Human Genetic Inventions, Patenting and Human Rights*, 2003, Canadian Ministry of Justice, at 47.

aside. We note that the HUGO's ethics committee '[i]n view of the ethical and logistical difficulties of defining community, … recommended that benefits be distributed broadly, perhaps to the health infrastructures of entire nations.'[46] This book will thus focus, instead, on another, more global, aspect of benefit sharing: as a tool to realise distributive justice in health.

Indeed, I will suggest that benefit sharing can be better justified under a theory of global distributive justice. This understanding of benefit sharing involves sharing outcomes with individuals and groups more generally, without having to refer to compensatory principles. As we will see in more detail in the course of this book, this obligation is based on the idea that justice requires us to protect the neediest and the most vulnerable and that mechanisms for ensuring transfer and assistance are required to further this goal on a global scale.[47] The duties imposed by global justice demand that individuals become involved in developing just global institutions and in supporting just domestic policies that affect individuals within and outside a nation's borders.[48] Therefore, the benefits arising from genetics should be distributed in a way that contributes to everyone's equality of opportunity and benefits the least well-off.[49]

I will also argue that the particularities of genetics also necessitate a global rather than individual perspective on benefit sharing in this area of research. For example, the fact that the human genome has symbolically been qualified as the *common heritage of humanity*[50] highlights its universal value for the human race and serves as a reminder that knowledge about the human genome should benefit humanity as a whole (including future generations) instead of serving narrow economic interests.[51] Moreover, the concept of common heritage associated with the human genome involves a notion of solidarity

---

46 HUGO Ethics Committee, 'HUGO Urges Genetics Benefit-Sharing' (2000) 3 *Community Genet* 88, at 90.

47 C. Beitz, 'Social and Cosmopolitan Liberalism' (1999) 75:3 *International Affairs* 515 at 518.

48 C.R. Beitz, 'International Liberalism and Distributive Justice: A Survey of Recent Thought' (1999) 51:2 *World Politics* 269, at 278 and 280.

49 J. Rawls, *A Theory of Justice* (Cambridge, Harvard University Press, 1971) at 7–8 (difference principle); J. Rawls, *A Theory of Justice*, 2nd edn, (Oxford: Oxford University Press, 1999) at 63, 72–73; C. Beitz, 'Rawls's Law of Peoples' (July 2000) *Ethics* 7; A. Buchanan, 'Rawls's Law of Peoples: Rules for a Vanished Westphalian World'(July 2000) *Ethics* 697.

50 UNESCO *Universal Declaration on Human Rights and the Human Genome,* Paris, 1997, art. 1.

51 C. Joyner, 'Legal Implications of the Concept of the Common Heritage of Mankind' (January 1986) 35:1 *International and Comparative Law Quarterly* 190; see also UNESCO, *International Consultation on the outline of the Universal Declaration on the Human Genome, Summary of the Response to the Questionnaire*, 1997; HUGO,

based on the fact that we share our genetic makeup (99.9%) with all other human beings, that it is 'part of every individual and integral to the evolution of the human species'.[52] However, the reality is that genetic applications remain inaccessible to many individuals all over the world because they do not have access to sufficient financial, infrastructure, and human resources to make use of this precious knowledge. As Thorsteindottir et al. clearly state, '[g]enomics is only a public good to those countries that have the capacity to exploit genomics knowledge and to conduct genomics research. Because of the need for these "access goods", genomics becomes a "club good", accessible mainly to industrialised countries.'[53]

This is enough to be deeply concerned about the way current assumptions define what is just and acceptable with regard to global access and distribution of resources in this field. It also highlights the need for a precise and enforceable concept of global benefit sharing in health that would position genetics as an essential tool for achieving global health rather than as a luxury beyond the reach of the most vulnerable people. To this end, we need to establish a normative basis for undertaking benefit sharing with developing countries in the global health sphere. Global benefit-sharing obligations can be built on theoretical and legal grounds but have not, up to now, been elaborated at length in the field of health and genetics. This will be the core of my thesis. I will flesh out the widely-used concept of benefit sharing to determine how we could ensure that it is used to further global health without leaving the most vulnerable out of the process.

THE STRUCTURE OF THE BOOK

Given the importance of genetics to human health on a global scale (as discussed below), this book will assess two important legal regimes – intellectual property and international human rights – to determine to what extent they further the goal of distributing the benefits arising from genetics equitably and globally. This evaluation is vital to ensure that legal regimes assist in ensuring that this promising field develops in a way that improves global health without leaving the most vulnerable outside of the process. Such an

---

Statement on Benefit Sharing, *supra* note 37, *Common Heritage*: 'While not respected by all nations, the concept of common heritage also resonates under international law (e.g. the sea, the air, space, etc.). Applied to human genetics, it maintains that beyond the individual, the family, or the population, there is a common shared interest in the genetic heritage of mankind.'

[52] C. Joyner, op cit, at 194.

[53] H. Thorsteinsdottir et al., 'Genomics, a Global Public Good' (2003) 361 *Lancet* 891 at 892.

investigation has not, so far, been developed at length with respect to the field of global health and genetics. I will undertake this complex task by employing and building upon cosmopolitan liberal theories of distributive justice developed over the few last decades as an extension of the work of Rawls and Daniels.

The first purpose of this book is thus to set out a grounding theory or theoretical framework to justify engaging in a global and more equitable redistribution of benefits produced by genetics. Ultimately, my analysis will produce strong normative benchmarks based on justice considerations that take needs into account rather than market-based power when evaluating major social, political, and legal implications resulting from the commercialisation of genetics.

The first chapter will set the contextual basis of my framework by providing justifications for a global application of distributive justice principles. To assess institutions and practices, I will propose a cosmopolitan methodology based on a global scheme of cooperation emerging from the idea of the universal importance of every human being as a unit of moral concern. This exercise will give us a sense of how institutions involved in the distribution of genetic benefits should function and within which specific parameters they should handle distribution.

The second chapter of this first theoretical part will elaborate an ideal conception of distributive justice in health to justify global access to genetics. I will establish normative grounds as the basis for my scheme of global health/health care justice, focusing on the special characteristics of health and on its crucial role in normal human functioning. After arguing that health is a crucial element of normal functioning, I will analyse the impact of normal functioning on the lives of individuals, using the criterion of the range of normal opportunities available to people. This will help us establish clear links between health problems, lack of access to the resources emerging from genetic research, and a diminution of the range of opportunities for which individuals of equal skill can build life plans. This discussion will highlight the specificity and universal importance of health. It will also flesh out my argument in favour of compensation for deviations from normal functioning and for the eradication of health inequities over which we can have some form of control through distributive justice mechanisms.

After this first part, I will have established a global distributive justice framework as the basis of my argument for more equitable and global access to health and genetics. The second part of the book will attempt to determine how and if my theory of distribution translates into positive law and to identify and analyse the main obstacles to legal compliance with global distributive justice. Although the development of genetics can affect many areas of law including privacy, employment, insurance, and criminal law, I will focus on two of the major international legal systems most concerned with distribution issues: intellectual property (IP) law (especially patent law) and human rights law. The first two chapters of the second part will be dedicated to the presentation and analysis of those international normative systems in order to determine if their underlying philosophy, structure, and functioning take account of the principles highlighted in my theoretical framework.

My analysis will conclude that these two legal frameworks regulating the distribution of benefits and resources arising from genetics are deficient, each in their own way, in the reach, operation, and substantive content of the standards they promote. Indeed, we will realise that, despite my argument for the universal special importance of health, this does not always receive the special and universal treatment it deserves in practice. The discussion will bring to light major power imbalances and a lack of focus on distributive justice issues mainly attributable to the political and economic contexts of application of the two systems and not to an irremediable incompatibility of the principles with diffusion and equitable access to knowledge. We will indeed realise that both systems, although very different in their nature and purposes, are driven mainly by market considerations either in their philosophy, principles, and/or application and that they do not give enough attention and importance to justice and solidarity issues. My analysis will bring us to acknowledge that the international order under which IP and human rights evolve inspires power struggles that shift our attention away from justice principles standing at the source of a shared morality and a cosmopolitan perception of humanity. My work will aim to highlight, analyse, and explain this reality.

Following my discussion on the conceptual link existing both between IP law and access and human rights law and access, my last chapter will focus on introducing practical examples to illustrate the intersection of IP and human rights law. Referring to a few examples, this last chapter will seek to highlight the practical impact that those two systems have had on scientific data-sharing and on availability and affordability of genetics research tools, products, and services in developing countries. Following the presentation of those examples, I will conclude this last chapter with a brief analysis of the intersection between IP rights and human rights in health. This will allow us to address the effects of strong and broad IP rights on the realisation and implementation of human rights and the tension existing between the two systems, both in terms of philosophy and application.

This will conclude the second and last part of this book dedicated to the assessment of the two major systems – first, with justice benchmarks established in the first theoretical part and secondly, with practical examples. Coming back to the evidence presented at the beginning of this work on the real potential of genetic research to improve global health, and on support for a notion of global distributive justice in health, we will be forced to realise that, as they currently function, the intellectual property and the human rights systems are not adequate to realise global benefit sharing in the field of genetics. Without arguing for the abolition of these systems or establishing detailed solutions and practical policy options, I will conclude the book with some suggestions of avenues that could be explored further to remedy this situation in order to further global distributive justice. This will set out a basis for further discussion on how we could work around some of the major obstacles identified throughout my analysis. It will also help us move from the vague and often symbolic ideal of benefit sharing actually prevailing toward the establishment of a real, enforceable concept of global benefit sharing in health that would position genetics at the rank of an essential tool for achieving global health.

PART I

A theoretical framework for distribution in health

The first part of the book examines the theoretical basis to ground my argument for engaging in a global and more equitable distribution of the benefits likely to emerge from genetic science based on health needs. This exercise is necessary to launch discussions and stir up debates on a common vision of the good and related universal basic needs, rights, and duties in order to establish appropriate principles of distributive justice in health. This framework will represent an ideal conception of global justice in health, a standard for appraising institutions and for guiding the overall direction of social change by providing a long-term goal of political endeavour and giving meaning to what we can do today in actual, existing conditions.[1] This analysis is very important although ideal principles of justice cannot always apply automatically and immediately to the practical reality. Indeed, we need to understand what we are compromising by accepting non-ideal conditions, and to receive guidance as to what we should be aiming for with respect to future social and institutional reforms. As clearly put by Schrecker: 'responsible ethical analysis must not regard crucial background elements of the social and economic context [...] as too big to change'.[2] To this end, a sense of moral responsibility for the actual state of the world must be developed and cultivated in order for reforms

---

1 J. Rawls, *A Theory of Justice*, *supra* chapter 1, note 49 at 128; C. Brown, *Sovereignty, Rights and Justice, International Political Theory Today* (Cambridge: Polity Press, 2002), at p. 180.

2 T. Schrecker, 'Benefit-Sharing in the New Genomic Marketplace: Expanding the Ethical Frame of Reference' in B.M. Knoppers (ed), *Populations and Genetics: Legal and Socio-Ethical Perspectives* (Leiden: Martinus Nijhoff, 2003).

and changes to gradually take place when they are indeed possible.[3] This part of the book will contribute to establishing the analytical basis required to lay the foundation for this important process.

I have split this first part into two main chapters. In the first, I will present an argument for a global application of justice principles, referring to a cosmopolitan approach that considers each human being with his or her basic health needs as a unit of consideration deserving equal attention. This global focus will give us the perspective we need to determine how institutions should work towards distribution of genetic-research benefits and who should be entitled to profit from this distribution.

The second chapter will establish an ideal scheme of global health/health care justice. To this end, we will need to reflect on the special importance of health, on the role it has to play in ensuring normal functioning and in the pursuit of an ideal of equality of opportunity for all. In this moral scheme, every individual's health interests receive equal consideration and the benefits arising from genetics are distributed so as to prevent health standards' differences caused by socio-economic factors. Our analysis will highlight the importance of compensating for the divergence from normal functioning and health inequalities over which we have some power through distributive justice schemes.

---

3 T.W. Pogge, 'The Moral Demands of Global Justice' (Fall 2003) *Dissent* 37.

1. Global application of distributive justice: a cosmopolitan approach

When choices are to be made regarding the ends and means of political action, or the structures and rules of institutions and practices, it is natural to ask by what principles such choices should be guided.[1]

INTRODUCTION

As discussed in the introduction, there is evidence that genetic research has and will continue to evolve to have significant positive effects on the health and lives of the people it reaches.[2] However, for the moment, we can expect that most genetic technologies will likely reach and benefit a limited number of people worldwide, the majority in developed countries. In fact, expensive innovations will probably continue to develop to address the needs of the affluent where there is a market for them and, in any case, will likely be accessible only to those people who have insurance coverage (public or private) or who can afford to purchase such technology with private funds.[3] Genetic discoveries could thus contribute to widen the health gap between rich and poor, both within and between countries, adding to the substantial inequalities that already characterise some health care systems and the global health agenda.[4]

---

1 C.R. Beitz, *Political Theory and International Relations* (Princeton: Princeton University Press, 1999) at p. 5.

2 See references cited in the Introduction, and see also Collins et al., 'A vision for the future of genomics research' (2003) 422 *Nature* 835; J. Bell, 'The Double Helix in Clinical Practice' (2003) 421 *Nature* 414; Program in Applied Ethics and Biotechnology and Canadian Program on Genomics and Global Health (University of Toronto Joint Center for Bioethics) *Top 10 Biotechnologies for Improving Health in Developing Countries*, Toronto, 2003; WHO, *Genomics and World Health*, Geneva, 2002.

3 M.J. Mehlman and J.R. Botkin, *Access to the Genome: The Challenge to Equality* (Washington DC: Georgetown University Press, 1998).

4 M. Leonard, 'Just Genetics: A problem Agenda' in T.F. Murphy and M.A. Lappé (eds), *Justice and the Human Genome Project* (California: University of California Press, 1994) 133.

Building upon my initial remarks on benefit sharing in the Introduction and in light of the concept's growing rhetorical importance in the field of genetics, it is now time to investigate its normative basis. In this chapter, we shall see that there exists a positive obligation to ensure equitable access to genetic advancements, that research priorities should be established accordingly, and that the benefits of genetics (knowledge, expertise, research tools, products and services, and profits) should be distributed more equitably globally, based on actual needs rather than simply on market forces.

1.1 DISTRIBUTIVE JUSTICE

1.1.1 What are the Characteristics of a Normative Obligation of Benefit Sharing?

Even if an ideal theory can seem far removed from the imperfect reality of developing countries, such a theory is essential to establishing the basis for concrete changes through a global normative framework for engaging in the international redistribution of resources produced by genetic and genomic research.[5] Indeed, theorists have developed different theories of justice to justify how goods, welfare, and services should be divided in a society.[6]

Liberalism is one of those theories widely applied to justice issues in political philosophy. One of the many components of liberals' ideal structure of

---

5 A. Kupler, 'Debate: Global Poverty Relief, More than Charity: Cosmopolitan Alternative to the *Singer Solution*' (2002) 16:1 *Ethics and International Affairs* 107.

6 For a good overview of the main theories of justice refer to: W. Kymlicka, *Contemporary Political Philosophy: An Introduction*, 2nd edn (Oxford: Oxford University Press, 2002). More specifically, on libertarianism, see: T.H. Engelhardt Jr, *The Foundations of Bioethics* (New York: Oxford University Press, 1986) at 342–343, R. Nozick, *Anarchy, State and Utopia* (New York: Basic Books, 1974), and for some critics of the application of libertarianism to health, see: L.M. Fleck, 'Just Health Care (I): Is Beneficence Enough?' (1989) 10 *Theoretical Medicine* 167; R.A. Epstein, 'Why is Health Care Special?' (1993) 40 *U. Kan. L. Rev* 307. For a new vision of libertarianism, one combining principles of libertarianism and of cosmopolitanism, see: D. Elkins, 'Responding to Rawls: Toward a consistent and supportable theory of distributive justice' (2007) 21 *BYU Journal of Public Law* 267.

Also, on utilitarianism refer to: F. Peter and T. Evans, 'Ethical Dimensions of Health Equity' in T. Evans et al. (eds), *Challenging Inequities in Health, from Ethics to Action* (Oxford: Oxford University Press, 2001) p. 25 at p. 28, and B. Williams, 'A Critique of Utilitarianism' in J.J.C. Smart (ed), *Utilitarianism: For and Against* (Cambridge: Cambridge University Press, 1973) p. 75. For a recent vision of how utilitarianism is compatible with principles of distributive justice ('distributive-sensitive justice'), see J. Schroth, 'Distributive justice and welfarism in utilitarianism' (2008) 51:2 *Inquiry* 123.

society is justice in goods and services through distribution to create more equitable circumstances. This is called distributive justice. Distributive justice is thus a perspective from which to consider justice in health matters and will be our focus in the establishment of our theoretical framework for global health distribution. Distributive justice aims at determining the equitable allocation and access to benefits (resources, services, goods) and burdens produced by social cooperation.[7] The most popular theory of distributive justice in the last 50 years was established by John Rawls in his book *A Theory of Justice*.[8] Unlike utilitarianism, distributive justice maintains that individuals have rights that cannot be sacrificed simply to create more benefits for others. In this view, social primary goods like liberty, opportunity, income, and wealth are to be distributed equally unless an unequal distribution will advantage the least well-off.

Similarly to other perspectives on justice like libertarianism and utilitarianism, principles of distributive justice are designed to allocate goods, resources and services when needs are greater than availability. They are however very different because they require equity and consideration of the most vulnerable in distribution, which make their application to health resources and technologies very relevant. The principles of distributive justice can differ according to the subject of the distribution (income, wealth, opportunities to the good life and so on), the beneficiaries (individuals, groups of persons, compatriots, foreigners and so on), the providers of the goods and services to be distributed (individuals, fellow citizens, governments, international organisations and so on), and the basis for the distribution (according to equality, to individual characteristics, to need and so on).[9] Within a theory of distributive justice, the role of luck, chance, and choice are also relevant when deciding distribution issues.[10] All of these characteristics allow us to consider health needs as a priority, and they are going to prove crucial for the examination of issues of global distribution in genetics.

1.1.2 Global Distributive Justice

Most theories of justice apply to domestic situations without dealing with the

---

7 C.R. Beitz, *Political Theory and International Relations*, *supra*, note 1.

8 J. Rawls, *supra* Introduction, note 49.

9 J. Lamont, 'Distributive Justice' (Fall 2003 Edition) *The Stanford Encyclopedia of Philosophy*, E.N. Zalta (ed), http://plato.stanford.edu/archives/fall2003/entries/justice-distributive (accessed: 26 January 2009).

10 C. Jones, *Global Justice, Defending Cosmopolitanism* (Oxford: Oxford University Press, 1999).

requirements of international distributive justice.[11] Nevertheless, if the benefits and burdens arising from social cooperation are the basis of distributive justice, economic interdependence at the international level and direction needed to make choices that can influence the well-being of individuals located in other societies might justify standards of global distributive justice analogous to the principles applicable within domestic societies.[12]

1.2 COSMOPOLITANISM: A WAY OF ENVISIONING GLOBAL JUSTICE

Current scientific developments that can help improve health and cure disease are universally essential and therefore should be accessible to all human beings who can physiologically benefit from them, by simple virtue of their humanity and needs. Cosmopolitanism provides a good starting point for the theoretical basis for such a premise.

There are two main categories of cosmopolitanism. First, institutional cosmopolitanism focuses on how political institutions should be established.[13] It holds that states and other political institutions should be restructured and placed under the control of an organisation akin to a 'world government' or other supranational political arrangement, so that we could see the world as a single entity in which individuals would be *citizens of the world.*[14] By contrast, moral cosmopolitanism focuses on the theoretical basis for the justification of institutions, practices, and interpersonal relations. I focus on this second type of cosmopolitanism and agree that human beings belong to one single community regardless of the presence of political institutional arrangements acknowledging this reality.[15]

---

11 For example J. Rawls, *The Law of Peoples* (Cambridge: Harvard University Press, 1999) at pp. 106, 114–119; J. Rawls, *supra* Introduction, note 49; M. Walzer, *Spheres of Justice* (Oxford: Blackwell, 1983). However, more recently, philosophers like Charles Beitz and Thomas Pogge have provided interesting arguments in favour of international distributive justice.

12 C.R. Beitz, 'International Liberalism and Distributive Justice', *supra* Introduction, note 48. Also see R. Pevnick, 'Political Coercion and the Scope of Distributive Justice' (2008) 56 *Political Studies* at 400.

13 As clearly explained by A. Miklos in 'Institutions in Cosmopolitan Justice' (2006) 20:3 *Global Society* 239.

14 D. Laertius, *Diogenes, in Lives of Eminent Philosophers,* trans. R.D. Hicks, Loeb Classical Library (London: William Heinemann, 1925) vol. 2, vol. 6:63.

15 This being said, the responsibility of ensuring that practices and rules are enacted in compliance with this moral cosmopolitan ideal should fall on institutions.

Moral cosmopolitanism is not associated with any specific political programme or philosophical theory, but is instead characterised by its perception of the moral basis upon which justice issues should be evaluated and of the proper scope of moral principles.[16] It establishes conditions that any acceptable approach to justice ought to meet. Cosmopolitanism does not demand specific measures. Instead, measures would flow from a specific theory of justice like global distributive justice.[17] Once a cosmopolitan direction is established, specific justice measures focus on different aspects of importance to each individual, including subjective elements like happiness, well-being, desire, and preference, as well as objective factors like needs, abilities, and opportunities.[18]

A moral cosmopolitan viewpoint is impartial, universal, individualist, and egalitarian in nature.[19] For cosmopolites, individuals are the fundamental entities of moral concern, as expressed by Thomas Pogge: 'every human being has a global stature as the ultimate unit of moral concern'.[20] Cosmopolites envision the social world as composed of persons rather than collectives.[21] They insist that each and every human being affected by institutional arrangements – like, for example, policy choices about production and distribution of burdens and benefits, or choices regarding the establishment of a specific institution – should be respected and given equal and impartial consideration by everyone, wherever they may be.[22] This perspective is based on the premise that individuals are entitled to certain treatment and consideration due to their humanity as opposed to other particularities including culture, politics, religion, and citizenship.

Human beings share a common sense of morality and common human interests in certain crucial spheres of universal importance. I agree with Buchanan when he says that: 'we should expect some congruence of moral

---

[16] O. O'Neill, *Towards Justice and Virtue* (Cambridge: Cambridge University Press, 1996) at p. 172; O. O'Neill, *Bounds of Justice* (Cambridge: Cambridge University Press, 2000) chapter 10.

[17] C.R. Beitz, 'Social and Cosmopolitanism Liberalism', *supra* Introduction, note 47 at p. 515; C.R. Beitz, *Political Theory and International Relations*, *supra* note 1; A Kupler, 'Rawlsian Global Justice, Beyond the Law of Peoples to a Cosmopolitan Law of Persons' (October 2000) 28:5 *Political Theory* 640.

[18] T.W. Pogge, 'Cosmopolitanism and Sovereignty' in C. Brown (ed), *Political Restructuring in Europe, Ethical Perspectives* (NewYork: Routledge, 1994) p. 89.

[19] C. Jones, *supra* note 10.

[20] T.W. Pogge, 'Cosmopolitanism and Sovereignty' (October 1992) 103 *Ethics* 49.

[21] C.R. Beitz, 'Rawls's Law of Peoples', *supra* Introduction, note 49.

[22] O. O'Neill, 'Hunger, Needs and Rights' in S. Luper-Foy (ed), *Problems of International Justice* (London: Westview, 1988).

values across societies, given the roles that morality plays in human life [...]'.[23] Those principles and values are the ones that play a role in preventing people from being exposed to serious harm and allowing them to pursue decent human lives through access to an appropriate range of opportunities.[24] Access to health and genetic resources is an example where the interests at stake could be so universally crucial that extending the range of rights and obligations beyond the level of citizenship is justified. The universalistic focus of moral cosmopolitanism can thus be justified by the common characteristics shared by all individuals, especially when we talk about genetics. Indeed, human beings share a similar genetic makeup and are physiologically alike.[25] Our common genetic heritage thus transcends geopolitical borders. Consequently, a number of genetic research projects that aim to identify significant genetic variations and to determine who such variants affect, have been undertaken worldwide. Such studies generate important information for the screening of individuals, families, and populations more genetically at risk or susceptible to certain diseases and conditions.[26] Therefore, even if different individuals may end up being personally affected by genetic discoveries in very different ways, at present it is difficult to predict who may benefit the most. Due to the similar characteristics we all share as human beings, it is safe to say that genetic developments carry a potential to benefit many individuals worldwide, both from a global community perspective and from a personal and familial perspective, regardless of the direct practical outcomes that might emerge from them, whether in the short or long term.

Genetics has an important collective aspect. In fact, we often refer to susceptible populations or at-risk groups; in some cases, reference has been made to the concept of *genetic nationalism*.[27] This group reference can have many different applications in genetics. Sometimes, the population aspect is not necessarily associated with specific diseases, conditions, or susceptibilities but is instead associated with existing boundaries as a practical, scientifically

---

23 A. Buchanan, *supra* Introduction, note 22 at p. 79.

24 S. Hampshire, *Innocence and Experience* (Cambridge: Harvard University Press, 1989) at 90.

25 Human Genome Project Information, *supra* Introduction, note 13.

26 A.J.F. Griffiths et al., *An Introduction to Genetic Analysis* (New York: W H Freeman and Co, 1999).

27 This attitude towards populations' genetic heritage has been observed in Iceland where the population has been presented with the idea that Icelanders are genetically special, that they might have some special genes and genetic conditions that cannot be observed elsewhere. H. Rose, *The Commodification of Bioinformation: The Icelandic Health Sector Database* (London: The Wellcome Trust, 2001) at p. 12.

relevant, and sometimes economically advantageous way to create a fixed heterogeneous genetic pool for research.[28]

In other cases, the group aspect of genetics is not at all clearly associated with existing political and geographical boundaries. In fact, susceptible populations will also be found in specific regions of the world, not necessarily clearly delimited, but associated for example with ethnicity, geographical distance from Africa or types of communities such as indigenous and tribal groups.[29] By contrast, other genetically at-risk populations are dispersed all around the world.[30]

---

28 Some of those genetic pools represent whole countries, for example, in Iceland where the government granted a 12-year licence to the company deCode to construct and operate a national health services database to link anonymous genotypes with medical records of consenting members of the population. Another national project is the UK Biobank project, a joint initiative from Wellcome Trust and the Medical Research Council. This project aims to recruit up to 500 000 men and women aged 45–69 from the general population across England, Scotland and Wales and use their blood samples, lifestyle details, and medical histories to create a national database to study the role of genetics and environmental factors in health and disease. Other initiatives are aimed at studying smaller populations, sometimes more homogeneous and isolated, like Sardinia and Israel. J. Kaiser, 'Biobank: Population Databases Boom, from Iceland to the U.S.' (November 2002) 298:5596 *Science* 1158; UK Biobank official website: http://www.ukbiobank.ac.uk; J.F. Merz, G.E. McGee and P. Sankar, 'Iceland Inc.?: On the Ethics of Commercial Population Genomics' (March 2004) 58:6 *Soc Sci Med.* 1201; S. Shifman and A. Darvasi, 'The Value of Isolated Populations' (2001) 28 *Nature Medicine* 309; C. Bourgain et al., 'Search for Multifactorial Disease Suceptibility Genes in Founder Populations' (2000) 64 *Annals Human Genetics* 255; R. Lampis et al., 'The Distribution of HLA Class II Haplotypes Reveals that the Sardinian Population is Genetically Differentiated from the Other Caucasian Populations' (2000) 56 *Tissue Antigens* 515; H. Lahat et al., 'A Missense Mutation in a Highly Conserved Region of *CASQ2* is Associated with Autosomal Recessive Catecholamine-Induced Polymorphic Ventricular Tachycardia in Bedouin Families from Israel' (2001) 69 *American J. Human Genetics* 1378.

29 This was in fact the purpose of the Human Genome Diversity Project established in 1993 to describe and understand the 1% difference and diversity in human genomes illustrated by many individual and population level differences. It aimed to collect biological samples from different population groups throughout the world, with the intention of building a representative database of human genetic diversity. It caused violent reactions from many of the indigenous groups targeted by the study, which gave rise to a project review by the US National Research Council in 1997. Since April 2002, a collection of more than 1000 DNA samples from 521 populations representing most of the world's genome variation has been available to non-profit research laboratories through collaboration between the HGDP and the Fondation Jean Dausset-CEPH in Paris. The HapMap project is a similar initiative. It seeks to understand the basis of genomic variation among unaffected individuals of similar ancestry to affected individuals by the identification of the genetic components of complex diseases and of variation in response to environmental exposures and to drugs. The long-term goal of

In any case, the majority of the potentially susceptible groups and populations still need to be identified. To do so, extensive genetic research must be undertaken and carried out globally, sometimes with no clear focus on specific populations and without restrictions based on borders. Indeed, even if the vital and obvious importance of the group and population aspect in genetics is acknowledged – especially in research[31] – we need to broaden our focus in taking the health needs of individuals into account. In fact, it is my contention that a clear focus on the needs of specific populations would be too narrow for the purpose of the global justice framework required for genetics. A univer-

---

the International HapMap Project, a collaborative endeavour among scientists in Japan, the UK, Canada, China, Nigeria, and the US, is to develop a haplotype map of the human genome that could shed light on the common patterns of human DNA sequence variation. Recent findings report that although linkage disequilibrium is different from one population to another they often share similar haplotype structure. D.F. Conrad et al., 'A worldwide survey of haplotype variation and linkage disequilibrium in the human genome' (2006) 38 *Nature Genetics* 1251. For more on these initiatives see: L.B. Barreiro et al., 'Natural selection has driven population differentiation in modern humans' (2008) 40 *Nature Genetics* 340; The International HapMap Consortium, 'A second generation human haplotype map of over 3.1 million SNPs' (18 October 2007) 449 *Nature* 851; I.G. Romero, 'How accurate is the current picture of human genetic variation?' (2009) 102 *Heredity* 120; M. Jakobsson, 'Genotype, haplotype and copy-number variation in worldwide human populations' (21 February 2008) 451 *Nature* 998; M. Dodson and R. Williamson 'Indigenous Peoples and the Morality of the Human Genome Diversity Project' (1999) 25 *Journal of Medical Ethics* 204; M.W. Foster, 'Integrating ethics and science in the International HapMap Project' (June 2004) 5:6 *Nature Reviews Genetics* 467.

30 For example, the genetic components of more common conditions like breast cancer and hypertension are being investigated in many centres, all around the world. D.H. Choi et al., 'Incidence of BRCA1 and BRCA2 Mutations in Young Korean Breast Cancer Patients' (May 2004) 22:9 *J Clin Oncol.* 638; S. Malander, 'One in 10 Ovarian Cancer Patients Carry Germ Line BRCA1 or BRCA2 Mutations: Results of a Prospective Study in Southern Sweden' (February 2004) 40:3 *Eur J Cancer* 422; B. Gorski et al., 'A High Proportion of Founder BRCA1 Mutations in Polish Breast Cancer Families' (July 2004) 110:5 *Int J Cancer* 683; N. Kato, 'Genetic Analysis in Human Hypertension' (May 2002) 25:3 *Hypertens Res.* 319; H.C. Hendrie et al., 'Alzheimer's Disease, Genes, and Environment: the Value of International Studies' (February 2004) 49:2 *Can J Psychiatry* 92.

31 Especially for population studies where there is much debate on the necessity and relevance of group consent and protection from potential harm due to improper disclosure. For more details: National Research Council (Committee on Human Genome Diversity), *Evaluating Human Genetic Diversity* (Washington DC: National Academy Press, 1997) at 4, 63–65, online NAP http://books.nap.edu/books/0309059313/html/index.html (accessed 4 June 2009); V. Arnason, 'Coding and Consent: Moral Challenges of the Database Project in Iceland' (2004) 18:1 *Bioethics* 27; M.J. Smith, 'Population-based Genetic Studies: Informed Consent and Confidentiality' (December 2001) 18:1 *Santa Clara Comput High Technol Law J.* 57.

salistic approach is more appropriate and compatible with a moral cosmopolitan viewpoint.

Another aspect of universalism relates to the fact that moral cosmopolitanism is not convinced that boundaries between territorial and political structures should have much moral importance.[32] Principles of justice should apply to the *global community of world citizens*,[33] those who live in different countries and with whom we can seem to share little in terms of culture, language, and customs, for example.[34] The cosmopolitan perspective requires scepticism about strong nationalism and patriotism when they have the effect of prioritising only social and political affiliation in the provision and distribution of aid. It calls instead for a sense of community among human beings in a universal comity of nations where borders are less significant.[35] However, this does not mean that cosmopolitanism is indifferent to local poverty and deprivation, as some suggest.[36] Instead, cosmopolitanism holds that the state level should not be given absolute priority when considering justice. As such, we should care about the focus of our distributive justice obligations, such as deprivation and pain, wherever they exist. In fact, even if nations are an important part of the existing political picture (and cosmopolitans are not necessarily arguing for their abolition, as discussed below), the moral significance of boundaries should be justified in terms of the values and ethical principles that are chosen and the priorities such a choice represents for every individual affected.[37] In other words, nationality-based special treatments and group loyalty are appropriate but are necessarily complemented by concurrent moral obligations to

---

32 For more on the cosmopolitan arguments on advocating the belief that boundaries between territorial and political structures should not have much moral importance in the application of principles of justice, see: Darrel Mollendorf, 'Equality of Opportunity Globalized?' (2006) 19:2 *Can. J. L. Jurisprudence* 301, at 304; Joel P. Trachman, 'Welcome to Cosmopolis, World or Boundless Opportunity (2006) 39 *Cornell Int'l L.J.* 477, at 478, and Kor-Chor Tan, 'The Boundary of Justice and the Justice of Boundaries: Defending Global Egalitarianism' (2006) 19:2 *Can. J.L. and Jur.* 319, at 343–4.

33 W. Hinsch, 'Global Distributive Justice' (January 2001) 32 (1/2) *Methaphilosophy* 58.

34 W. Scheuerman, 'Globalization' *The Stanford Encyclopedia of Philosophy (Fall 2002 Edition)*, E.N. Zalta (ed), http://plato.stanford.edu/archives/fall2002/entries/globalization (accessed 6 May 2009).

35 I. Kant, *The Metaphysical Elements of Justice* (1797), trans. J. Ladd, 2nd edn (Indianapolis: Hackett Publishing Co., 1999).

36 G. Fletcher, *Loyalty: An Essay on the Morality of Relationships* (Oxford: Oxford University Press, 1993) at 21.

37 C.R. Beitz, 'Cosmopolitan Liberalism and the States System' in C. Brown (ed), *Political Restructuring in Europe, Ethical Perspectives* (NewYork: Routledge, 1994) chapter 6, p. 123, at 124.

individuals beyond our border. The priority given to our fellow citizens for distributive justice is not absolute; other human beings who may not be citizens can also have legitimate interests in distribution. The latter should be given important consideration in cases where the interests at stake are significant, as are, for example, claims for the protection of basic rights or vital interests.[38] This brings us to another important characteristic of cosmopolitanism: individualism.

The individualistic vision of the self, defended by cosmopolitans, is the object of much criticism, especially from proponents of a communitarian approach to justice. Indeed, communitarians critique a universal vision of justice and needs, instead arguing for variable principles of justice originating from societies' historical, institutional, and cultural particularities. They argue that distributive justice will only find logical application within restrained social groups who share subjective needs.[39] In the field of health, efforts to apply a single common morality and to adopt a universal approach when dealing with issues arising with the production and applications of science and medicine are criticised. Indeed, such attitude is seen as an effort from the western world to export its conception of what is ethical into an area where moral meanings of the most basic concepts like disease and health can differ between countries and religions.[40] The communitarian perspective of justice deserves great consideration and can be of considerable help in approaching and resolving the important justice debate in health-related matters.[41]

---

[38] C. Jones, *supra* note 10; P. Kleingeld, E. Brown, 'Cosmopolitanism' *The Stanford Encyclopedia of Philosophy (Fall 2003 Edition)*, E.N. Zalta (ed), http://plato.stanford.edu/archives/fall2002/entries/cosmopolitanism (accessed 24 January 2009); Mathias Risse, 'What to Say About the State', (February 2006) Harvard Kennedy School of Government Faculty, Research Working Paper Series, at 25.

[39] One of the arguments often put forward to support this view is that human needs are *socially relative* and that arguing for a universal standard of needs could give rise to attempts by the more powerful to impose their vision of needs and that this could result in *cultural imperialism*. For discussions on communitarianism and critique of the universal perspective of liberalism, see: M. Sandel, *Liberalism and the Limits of Justice*, 2nd edn (Cambridge: Cambridge University Press, 1998); C. Taylor, *Philosophy and the Human Sciences: Philosophical Papers 2* (Cambridge: Cambridge University Press, 1985) chapter 1; W. Kimlicka, *Contemporary Philosophy: An Introduction* (Oxford: Oxford University Press, 2002), chapter 6; P. Marshall and B. Koenig, 'Accounting for Culture in a Globalised Bioethics' (2004) 32 *Journal of Law, Medicine and Ethics* 252; A. Heller, *The Theory of Human Need in Marx* (London: Allison and Busby, 1976) at pp. 96–97; S. Scheffler, 'Conceptions of Cosmopolitanism' (1999) 11 *Utilitas* 255, at 256.

[40] P. Marshall and B. Koenig, op cit at 252 and 256; D. DeGrazia, 'Common Morality, Coherence, and the Principles of Biomedical Ethics' (2003) 13:3 *Kennedy Institute Journal of Ethics* 219.

[41] However, we have to be careful with any view that uses cultural differences

Although acknowledging and appreciating that different individuals and cultures can have different views of the definition and importance of health, in this book, I adopt a universal perspective on health. Even if different perceptions on some aspects of health coexist, I consider that health is something universally desirable, that it is a state of normal functioning influenced by numerous biological, genetic, socio-economic, psychological and environmental factors and which allows people to accomplish and further important life goals. In other words, I consider that health is an objective basic human need and consequently that access to health and genetics (as previously defined) should be an issue of universal importance for every human being no matter who they are, where they come from, or where they live.[42] I, of course, acknowledge that some identified groups might have specific vulnerabilities to disease and additional health needs, and that they should be able to be treated accordingly when they are identified. However, this does not mean that the basic health needs of individuals from those groups should be considered any differently from those of any other individual.[43] With this perspective in mind, I leave the communitarian perspective aside and adopt a moral cosmopolitan viewpoint to address the issues at stake. This being said, I appreciate that a culture-specific strategy and sensitivity could be essential, in the long run, to understand and address local and cultural specificities in the delivery of genetic products and services, and for an ethical provision of genetic counselling services, for example. However, this goes beyond the scope of this book.

On the practical level, some characterise the existing world order as a structure that institutionalises rather than eradicates oppression, self-interest, and deception on a global scale.[44] If valid, this clearly contradicts the universal conception of justice promulgated here. If we start with a cosmopolitan model of moral reciprocity in which all individuals are seen and treated as equals, we

---

to justify inequities and tolerate suffering. For a very good analysis and critique of such approach refer to P. Farmer, 'On Suffering and Social Violence: A View from Below' in A. Kleinman, V. Das and M. Lock (eds), *Social Suffering* (Berkeley: University of California Press, 1997) 278; for an illustration of this opinion, see also J. Mukherjee, 'HIV-1 Care in Resource-Poor Settings: A View from Haiti' (2003) 362 *Lancet* 994.

42 For more on the universal importance of certain needs, refer to the enlightening parallel established between the universality of basic needs and of suffering emerging from imperialism by Doyal, where he argues that, in both cases, victims' cultural background should not be used to assume that their suffering is qualitatively different. L. Doyal and I. Gough, *A Theory of Human Need* (New York: The Guildford Press, 1991), at pp. 29–30.

43 Ibid., at. p. 57.

44 O. O'Neill, *Face of Hunger: An Essay in Poverty, Development and Justice* (London: Allen and Unwin, 1986) at p. 145; K.-C. Tan, 'Kantian Ethics and Global Justice' (Spring 1997) 23:1 *Social Theory and Practice* 53.

need to address and eradicate the hierarchy and inequities present in the world.[45] In practice, this should happen through the promotion of a cosmopolitan institutional reform that would directly influence the choice and design of the norms that regulate property and cooperation. Such an institutional approach to moral cosmopolitanism requires that the world as a whole provide the context for determinations of justice[46] and that the criteria of distributive justice actually prevailing domestically be applied globally for the satisfaction of the just interests of all individuals. This certainly appears as a long-term goal in the actual global reality due to the limited enforcement capability of institutions on the international scene. However, this does not mean that we should not pursue such an ideal. As clearly stated by Buchanan:

> [a]lthough at present it is unrealistic to expect that the international legal order can do much directly to achieve distributive justice by formulating and implementing comprehensive principles of distributive justice, it is nonetheless an important element of the ideal moral theory of international law.[47]

This issue will be addressed all through this chapter as we assess the importance of an ideal theory.

1.2.1 Objections to Cosmopolitanism

As indicated in the previous section, a cosmopolitan view does not give absolute priority to compatriots for the distribution of certain goods and services; instead it focuses on the universal equality of individual needs, regardless of nationality or geographic location. In the last section, I have said a few words on the communitarians' arguments against cosmopolitanism but the most common and severe critiques of cosmopolitanism are proponents of the view that envisioning distributive justice in a cosmopolitan manner overlooks important elements of state sovereignty and autonomy and ignores the special relationship that prevails between people from the same community. In this section we will analyse and refute those two criticisms, arguing that the importance of access to health transcends boundaries and that the universal aspect of health is demonstrated by the existence of global interdependence.

---

45 K. Nielsen, 'Global Justice, Capitalism, and the Third World' in J. Arthur and W. H. Shaw (eds), *Justice and Economic Distribution* (Englewood Cliffs: Prentice Hall, 2001) at p. 236.

46 T.W. Pogge, 'Cosmopolitanism and Sovereignty' *supra* note 18, at 97; R. Forst, 'Toward a Critical Theory of Transnational Justice' (January 2001) 32:1/2 *Methaphilosophy* 26.

47 A. Buchanan, *supra* Introduction, note 30, at p. 203.

1.2.1.1 From the defenders of state sovereignty and autonomy

A common critique of cosmopolitanism is that it fails to adequately acknowledge the concept of state sovereignty. Many consider states to represent the principal independent ethical institutions in the world.[48] They are autonomous bodies that have the power to exercise control and enforce rights over their territory and over their citizens. The sovereignty of states is a basic principle of international law[49] which provides that all states are juridical equals, despite important differences in political and economic power. Consequently, states have the autonomy to set up their own domestic rules and exercise political coercion, but may concede part of their sovereignty by voluntarily agreeing to comply with international norms. Thus, often stimulated by self-interest, states can freely decide whether and how they choose to participate in the establishment and preservation of international norms that deal with issues arising beyond national boundaries and jurisdictions. Such a perspective exemplifies the concept of political freedom that accentuates the role of the nation-state and presupposes that every state is driven by its own national interests, such as preserving its political autonomy, its territorial integrity, and expanding its economic system.

However, this so-called realist vision of the state only represents one way of envisioning the role of nations. As they exist today, states lack unlimited sovereignty, notwithstanding any desire they may have to pursue their own interests in the creation or support of international obligations and institutions. International treaty obligations and the new existing global order confine nation-states to a more limited concept of sovereignty, the limits of which are partially determined by their respective political and economic positions. For example, nation-states may be restricted with respect to how they deal with other states, how they respect human rights, and how they enforce their international obligations depending on their political situation, their economic power and on the strategies adopted by other very powerful non-state actors.[50]

Advocates of state sovereignty accord significant ethical and moral weight to state boundaries and autonomy despite the cosmopolitan argument that they

---

[48] M. Frost, *Towards a Normative Theory of International Relations* (Cambridge: Cambridge University Press, 1986) at pp. 177–183; M. Frost, *Ethics in International Relations: A Constitutive Theory* (Cambridge: Cambridge University Press, 1996) at pp. 150–155. See also: A.E. Eckert, 'Obligations beyond national borders: international institutions and distributive justice' (2008) 4:1 *Journal of Global Ethics* 67. Eckert argues in this article that at the international level states are the most significant moral agents.

[49] This concept has been codified in the Charter of the United Nations, 892 *U.N.T.S.* 119, art. 2(1): 'The Organization is based on the principle of the sovereign equality of all its Members.'

[50] O. O'Neill, 'Agents of Justice' (January 2001) 32:1/2 *Metaphilosophy* 180.

are 'historically determined but morally arbitrary features of the earth's political geography'.[51] Some *statists* believe that each sovereign and autonomous state has a certain responsibility for any underdevelopment and poverty, arguing that such conditions are often directly related to internal, structural, and political problems and traditions. They believe that the cosmopolitan ideal, which considers the global context as the basis for justice, is utopian and would violate the limited but important degree of domestic, institutionalised social cooperation that some states have reached.[52] This leads some to believe that only states' citizens are entitled to be compensated for deprivation with the application of principles of distributive justice and economic egalitarianism. These same criteria do not apply on the international scene where they think that only some sort of minimal threshold of absolute deprivation should be compensated.[53] However, those opponents of cosmopolitanism do not seem to acknowledge the increasing economic and political interdependence among states at present, which causes states to lose part of their sovereignty due to globalisation.[54] Indeed, this is giving rise to an entirely new and sophisticated global order.

Another critique of cosmopolitanism originates from a sense of nationalism and is based on the idea that cosmopolitanism fails to recognise the value of individuals' rights and affiliations to their community as constituting a crucial part of the enjoyment and satisfaction of life.[55] Some, like Drahos, argue that even if we can observe interdependence between states in various sectors, it does not mean that those states are forming a system of mutual cooperation.[56] Indeed, various elements characterise nationality as compared to other sources of collective identity: the fact that nationality develops from a shared belief in its existence, distinct rules, cultural conceptions, and values; its origin in history; its connection to a specific geographic region; and its reflection in individuals' distinct and subjective identification.[57] Nations are thus viewed as

---

51 C. Beitz, 'International Justice: Conflict' in L.C. Becker and C. Becker (eds), *Encyclopedia of Ethics* (London: Garland, 1992) at p. 623.

52 C. Jones, *supra* note 10.

53 M. Blake, 'Distributive Justice, Sate Coercion and Autonomy' (2001) 3 *Philosophy and Public Affairs* 257, at 264.

54 E.R. Gold et al., 'The Unexamined Assumptions of Intellectual Property: Adopting an Evaluative Approach to Patenting Biotechnology Innovation' (October 2004) *Public Affairs Quarterly* 299.

55 C.R. Beitz, *Political Theory and International Relations*, *supra* note 4 at pp. 290–291.

56 P. Drahos, *A Philosophy of Intellectual Property* (Aldershot: Dartmouth Publishing Company, 1996) chapter 8, at pp. 170–198.

57 C. Taylor, *Reconciling the Solitudes* (Kingston and Montreal: McGill-Queens, 1993) chapter 3; D. Miller, 'The Nation-State: a Modest Defence' in C. Brown (ed), *Political Restructuring in Europe, Ethical Perspectives* (New York: Routledge, 1994) chapter 7, p. 136, at p. 141.

major sources of solidarity, crucial in circumscribing specific duties of aid, assistance, and support to other citizens and therefore helpful for domestic justice. Nationalism thus allows individuals to forge bonds as they share a similar sense of identity[58] and is also often perceived to be the guardian of distinct cultures that may not be recognised at the global level.[59] It therefore rejects the idea of a world group to which duties of distributive justice can be applied. Nevertheless, it is worth noting that cosmopolites do not necessarily argue for the abolition of states. What they oppose are boundaries that systematically inflict injustices on outsiders and the existence of restrictive domestic welfare schemes and citizenship rights 'held by persons *qua* citizens rather than directly or exclusively *qua* human beings'.[60]

A good illustration of an ideology that would counter cosmopolitanism is set forth by Rawls in his treatment of international relations and global justice obligations in his *Law of Peoples*, which includes some of the critiques described above.

1.2.1.2 Rawls' *Law of Peoples* and international justice

Rawls's *Theory of Justice* is one of the most well-known treatments of justice in the last 50 years. His theory established the principles of distributive justice, but they were applicable only to individual states, a circumscribed context where it would be possible to identify social cooperation from which rights and duties arise. In *The Law of Peoples*, Rawls offers an extension of his theory of justice beyond the individual state. Rawls changes his theoretical contract mechanism for the specific context of international justice and the parties become representatives of peoples, rather than individuals or persons, who make choices about terms of cooperation that are 'fair to peoples and not to individual persons'.[61] Individuals are not the relevant, moral players in the global setting since their distributive justice claims have already been taken into account at the domestic level, where justice principles are constructed independently from principles of global justice.[62]

---

58 C. Brown, *supra* Part I, note 1, at 180; O. O'Neill, 'Justice and Boundaries' in C. Brown (ed), *Political Restructuring in Europe: Ethical Perspectives* (London: Routledge, 1994) at p. 85.

59 J. Carens, 'Migration and Mortality: A Liberal Egalitarian Perspective' in B. Barry and R.E. Goodin (eds), *Free Movement* (Hemel Hempstead: Harvester Wheatsheaf, 1992) at p. 23.

60 D. Harris, *Justifying State Welfare: the New Right Versus the Old Left* (Oxford: B. Blackwell, 1987) at p. 147; S. Caney, 'Cosmopolitan Justice and Equalising Opportunities' (January 2001) 32:1/2 *Metaphilosophy* 113.

61 J. Rawls, *supra* note 11, at p. 17.

62 A. Kupler, *supra* note 5; A. Kupler, *supra* note 17.

For Rawls, peoples are self-contained societies, but are not necessarily liberal democratic societies.[63] This means that hierarchical societies can be considered as peoples (qualifying as well-ordered societies) for the application of the law of peoples, a notion which is problematic.[64] Rawls views peoples rather than states as the primary agents of justice at the international level. He argues that peoples differ from states in three fundamental respects: peoples do not have the right to go to war to further their interests, they must meet certain minimal standards in their internal affairs, and they are fully prepared to grant the very same respect and credit to other peoples as equals.[65] However, his description of peoples is very similar to the definition one would give of states:

> Liberal peoples do, however, have their fundamental interests as permitted by their conceptions of right and justice. They seek to protect their territory, to ensure the security and safety of their citizens, and to preserve their free political institutions and the liberties and free culture of their civil society.[66]

The equality project supported by Rawls on the global scene is a political equality of just or decent peoples, mainly structured as states, not an equality of persons as typified by cosmopolitanism. The main objective of Rawls' theory of international justice is to push societies to the point where it becomes possible for them to support just and decent institutions. The focus is not on the material comfort and well-being of persons individually, but more on achieving a world of peaceful and decent societies. In such a world, justice issues are not triggered by inequities between individuals who live in different regions of the world. The international redistribution Rawls calls for is not a consequence of the principles of cosmopolitan global distributive justice. It deals instead with the global institutional structure and with the political and economic effects it can have on states and on their ability to continue implementing their principles of national justice.[67] It arises from an ideal conception of an international order that would be composed of distinct and independent, decent and autonomous domestic societies that cooperate on the basis of a similar conception of international justice, which would include a

---

[63] For interesting discussions on Rawls's international theory of justice, refer to: F.R. Teson, 'The Rawlsian Theory of International Law' (1995) *Ethics and International Affairs* 79; T. Pogge, 'An Egalitarian Law of Peoples' (Summer 1994) 23 *Philosophy and Public Affairs* 211; D. Moellendorf, 'Constructing the Law of Peoples' (June 1996) *Pacific Philosophical Quarterly* 77.

[64] I will not analyse the scope of Rawls' definition of peoples here, but for an interesting discussion on this aspect, refer to F.R. Teson, op cit.

[65] J. Rawls, *supra* note 11, at pp. 25, 26 and 35.

[66] Ibid. at p. 29.

[67] P. Drahos, *supra* note 56, chapter 8.

duty of mutual aid.[68] In fact, Rawls argues for a simple duty of assistance toward burdened societies in cases of extreme emergency and to help them develop their economy and reach the minimum requisite standard of internal organisation for their basic development and satisfaction of their populations' interests.[69] Rawls refuses to transpose his domestic 'difference principle' (any inequities should be to the greatest advantage of the least well off persons) at the global level as he considers it unacceptable for certain peoples to bear the burden of decisions made by other peoples.[70]

As clearly explained by Buchanan, Rawls likely adopted such international principles instead of principles of global distributive justice due to the lack of institutions and resources to implement the latter principles at the global level and the insufficient consensus on the nature of ideal principles of justice that exist among different peoples.[71] Also, Rawls believes that most obstacles to a society's sustainable economic and social advancement involve its own internal structure, culture, and tradition rather than its natural resource endowments or position in the international political economy.[72] Such arguments require prioritisation of national citizenship and a focus on individual societies as opposed to individuals who live in the global international order. More importantly, they also call for a limited duty of assistance to burdened societies with a clear starting and cut-off point.

1.2.2 Reasons for Envisioning Distributive Justice on a Global Scale: A Response

Certainly, once we accept the case for a rights-based 'welfare state', we are, [...] morally constrained to go 'beyond the welfare state' to respect the same rights to optimum need-satisfaction on a global scale.[73]

In this next section, the limited *statist* focus of justice issues will be critiqued and compared to a scheme in which the principles of distributive justice are based upon global cooperation, which may emerge from increasing international globalisation and interdependence.

---

68 C.R. Beitz, 'Social and Cosmopolitan liberalism', *supra* Introduction, note 47; W. Hinsch, *supra* note 33.

69 J. Rawls, *supra* note 11, at p. 76.

70 J. Rawls, ibid. at pp. 118–120; T.W. Pogge, 'Moral Universalism and Global Economic Justice' (2002) 1:1 *Politics, Philosophy and Economics* 29.

71 A. Buchanan, 'Rawls's Law of Peoples: Rules for a Vanished Westphalian World', *supra* Introduction, note 49.

72 J. Rawls, *supra* note 11, at pp. 74–77 and 105.

73 L. Doyal and I. Gough, *supra* note 42, at p. 142 quoting G. Myrdal, *Beyond the Welfare State* (New Haven: Yale University Press, 1960).

1.2.2.1 The limits of boundaries and the reality of the global order

A cosmopolitan account of justice places us in front of the limits of the existing global order and its failure to consider every individual's interest. If we believe that each human being is entitled to equal consideration, the prevalence of such great poverty in the world must be considered problematic. Therefore, the assumption that international justice necessarily presupposes the existence of states – and is simply an additional topic to justice issues arising within isolated well-ordered societies that are delimited by clear boundaries – overlooks the actual state of the world. Instead of recognising each person as a unit of moral concern, rights and privileges are granted to people according to their citizenship and where they are geographically located. Nevertheless, the actual international socio-political situation clearly shows that many states are not only very ineffective at protecting justice within their boundaries, but often are also very unsuccessful at securing it outside of their territory. Hence, states are not necessarily the only or best actors to protect justice; indeed, it would be inaccurate to describe the current international environment as various states united by voluntary mutual-assistance endeavours that are promptly undertaken. If each human being has a right to be free from the suffering and indignities of poverty, it is unacceptable to contain redistribution within nation-state boundaries, as such a limited view threatens to leave many individuals from very poor societies in great deprivation.[74]

Our increasingly interdependent world is characterised by intense deterritorialisation, the spread of social relations across borders,[75] global capital and commodity markets,[76] and the rising power of multinational companies and other non-state authorities.[77] Therefore, foreigners are people with whom we do more and more business and trade in various sectors and with whom we collaborate and are involved in different economic, political, and cultural levels and settings. An unambiguous division between the national and international sectors becomes impossible and the vision of states as privileged actors in the realisation of normative ideals is also untenable. It is thus increasingly difficult to argue that considerations of distributive justice should be confined to existing state boundaries without referring to the broader context of their close connections with other foreign agents.[78] In other words, if we consider distant strangers as very involved in our politico-economic reality, we

---

74 C. Jones, *supra* note 10, chapter 7, at 173 et seq.

75 W. Scheuerman, 'Globalization' *The Stanford Encyclopedia of Philosophy* (Fall 2002 edition), E.N. Zalta (ed), http://plato.stanford.edu/archives/fall2002/entries/globalization (accessed 6 May 2009).

76 T.W. Pogge, 'Cosmopolitanism and Sovereignty' *supra* note 18, at 108.

77 A. Woods, *Hegel's Ethical Thought* (New York: Cambridge University Press, 1990) at p. 30.

78 S. Scheffler, *supra* note 39.

should not be entitled to adopt a different standard and ignore their presence when establishing a framework of distributive justice. Consequently, over-emphasis on the importance of state sovereignty in considering justice issues may obscure the reality of our interconnected and heterogeneous world: 'compatriots, intimates and kin, with whom we may share much, do not form an "ideal" united community that pre-empts plurality and the need for justice'.[79]

As Kant indicated, increasing international economic cooperation created, even 200 years ago, a new basis for international morality.[80] Indeed, the international picture is characterised by a basic global structure represented by various political and socio-economic rules and institutions involved in distributing burdens and benefits. This needs to be recognised in order to establish a theoretical framework for global distributive justice. The growing global disparity between rich and poor, increasing external control over domestic societies, and increasing global regulation and governance comprise a world that cannot be adequately restricted to the political theory of the nation-state.[81] States are active participants in this global structure, through the various international organisations and the states' respective influence over enforcement. As such, international organisations are responsible for the elaboration of various regional and international normative agreements, such as declarations and treaties on trade, human rights, and intellectual property and are at the centre of important collective initiatives, like international financial systems. Indeed, such organisations represent an important context for justice since their structural and normative characteristics influence the well-being of individuals and groups.

The elements of this international order play a crucial role in modelling the ways in which the burdens and benefits of the existing international cooperation scheme are distributed between countries and individuals.[82] There is an economic debate as to whether the international interdependence caused by globalisation has overall negative effects on the economic situation of the poor. Some have argued that globalisation widens the revenue and welfare gap between rich and poor countries and individuals of the world and, consequently, that the living conditions of the poor have deteriorated.[83] Others believe that global economic integration and the openness to international

---

79 O. O'Neill, 'Justice and Boundaries', *supra* note 58 at 74–82.

80 I. Kant, *supra* note 35, at pp. 124–129.

81 S. Hymer, *The Multinational Corporation: a Radical Approach* (Cambridge: Cambridge University Press, 1979) at p. 76.

82 For example, art. 28 of the Universal Declaration of Human Rights states that: 'everyone is entitled to a social and international order in which the rights and freedoms set forth in the Declaration can be fully realized'.

83 J.E. Stigliz, *Globalisation and its Discontents* (New York, W.W. Norton, 2002).

trade contribute to better income distribution, which decreases poverty rates and global inequality.[84] In both cases, there seems to be agreement that globalisation could only improve such conditions if complemented by distinct institutional and policy reforms. Therefore, it seems that the distributional consequences arising from growing interdependence and justness of international social, political, and economic arrangements must be assessed morally by a distinct theory of justice. Indeed, as explained by Buchanan:

> [...] because the workings of the global basic structure have such profound and enduring effects on individuals and groups – and because these effects are for the most part neither chosen nor consented to by those affected – the global basic structure is subject to assessment from the standpoint of justice.[85]

A theory of global distributive justice should be concerned with the basic organisation of international society – that is, the basic structure of political and economic power relations that influence the global distribution of burdens and benefits. While voluntary international transfer and assistance measures might be of some help in redressing certain inequities they are theoretically unjustifiable, unreliable, and unsupportive in terms of sustainable change.[86] What is required globally is a direct application of justice principles to the basic structure of the global regime to assess the moral character of international institutions.[87] This implies that justice might best be served by recognising that important institutions regulating human action do not all need to be

---

84 D. Dollar, 'The Poor Like Globalization but Institutions and Policies are Needed to Deliver the Hoped for Results' (23 June 2003) YaleGlobal, online: http://yaleglobal.yale.edu/display.article?id=1934 (accessed 26 January 2009); D. Dollar and K. Aart, 'Growth *Is* Good for the Poor', The World Bank Development Research Group, March 2000; World Bank, 'Global Poverty Down By Half Since 1981 But Progress Uneven As Economic Growth Eludes Many Countries' News Release 2004/309/S, 23 April 2004.

85 A. Buchanan, *Justice*, *supra* Introduction, note 30, chapter 2, at p. 76.

86 For example, see C. Jones, *supra* note 10; S. Scheffler, *supra* note 39.

87 C.R. Beitz, 'Social and Cosmopolitan liberalism', *supra* Introduction, note 47; C. Jones, ibid. See also T.W. Pogge, 'Cosmopolitanism and Sovereignty' *supra* note 18, at pp. 61 et seq, and T.W. Pogge, *World Poverty and Human Rights: Cosmopolitan Responsibilities and Reforms* (Cambridge: Polity Press, 2002), at pp. 168–195, where Thomas Pogge attempts to reconstruct the notion of sovereignty so that it may be compatible with a cosmopolitan vision and the existing global order. He suggests a multi-layered institutional scheme in which government authority is 'vertically dispersed' rather than concentrated almost completely within states, and where we find a number of political units governing individuals without hierarchy. This type of system would allow a shift from a domestic focus on justice to a global one, an essential step for moral cosmopolitanism. This is an interesting avenue for re-engineering the global political structure and it should be explored further.

territorially delimited.[88] Indeed, sovereignty should not be concentrated only at one level and may be differently envisioned, for example, as associated with functional tasks and non-territorial spheres of interaction.[89]

Together with the problems encountered by the application of strict concepts of sovereignty and territorial boundaries, another reason for envisioning justice at the global level is the lack of logical justification for adopting a double standard of justice.

1.2.2.2 The problem of a double standard of justice

Many injustices and inequities arise from the existing global order. If we rely on theories like Rawls' approach to international justice, it is possible for unjust international conditions to be neglected as international society is not held to the same standards of justice as domestic societies. How can this be justified? How can we, at the same time, qualify the major national inequalities as injustices, and yet find analogous inequalities morally acceptable in the global order?

The previous section posited the existence of a basic global structure shaped by complex international economic, political, and cultural relationships. This constitutes a global context for cooperation, analogous to its domestic counterpart and characterised by institutional inequities and justice violations. Such considerations are not acknowledged at all by Rawls, who clearly endorses two separate and distinct standards of justice at the domestic and the global levels.

An example of Rawls' application of different morality to the same reality can be found in his assessment of the distributional effects of a basic structure. At the national level, Rawls establishes that principles of distributive justice are vital to redressing inequalities created by basic national structures. At the global level, where an analogous basic structure exists, Rawls' theory does not transpose the difference principle, as he calls for a principle of charity in the form of an arbitrary duty to help burdened societies build decent political and social regimes, rather than establishing similar enforceable principles of justice.[90] He justifies this position on the ground that it is unacceptable for people to bear the burden of decisions made by other people.[91] What he does not say is why it is different when one province, township, family, or person bears the burden of decisions made by another at the national level.[92] If we

88 O. O'Neill, *supra* note 58.

89 Ibid.

90 Ibid.; J. Rawls, *supra* note 11, at p. 37.

91 J. Rawls, ibid. at pp. 118–120.

92 T.W. Pogge, *Realizing Rawls* (Ithaca: Cornell University Press, 1989) at pp. 253–253; T.W. Pogge, 'An Egalitarian Law of Peoples' (1994) 23 *Philosophy and Public Affairs* 195, at 211–213; T.W. Pogge, 'Moral Universalism and Global Economic Justice', *supra* note 70.

follow this reasoning, the imposition of a global economic order that generates great international inequality can be justified so long as the societies impoverished by this structure are charitably assisted to the extent of raising them above a certain minimum level.[93]

Rawls also argues that the great cultural differences between nations justify the different treatment of individuals from different nations and the establishment of domestic benchmarks of justice instead of global ones.[94] Even if different conceptions of domestic justice can lead to complex negotiations at the global level, some principles, such as equality in opportunities and democratic participation in the institutions of global governance, seem likely to receive a great deal of support from a majority of states. Furthermore, in practice, most countries' cultural and political traditions would likely not preclude preventive and curative genetic products and services from reaching their peoples. Indeed, in countries and communities where genetic testing and analysis are already used, there is evidence that they are widely accepted practices.[95]

It is crucial to reiterate that the ability of a society to reach a higher stage of development and address national inequalities at different levels is not only a result of domestic economic and political factors, but is also a direct consequence of structures and events beyond its borders.[96] Indeed, a society's participation in global political and economic relations can wreak havoc at the domestic level, on its social, economic, and political situation. In fact, rather than representing voluntary mutual-benefit schemes, the basic global structure is characterised by unequal bargaining power, mainly driven by a few powerful state actors seeking to advance their own economic interests.[97] When

---

93 T.W. Pogge, 'Rawls on International Justice' (2001) 21 *Philosophical Quarterly* 246.

94 J. Rawls, *supra* note 11, at p. 105.

95 Indeed, as reported by Marshall and Koenig, it appears to be a common practice within some orthodox Jewish communities to use early pre-conception genetic testing to restrain partners' selection and marriage options. Another example can be observed in Cyprus, where broad prenatal testing for thalasemia is imposed and subsidised by the government and very well accepted by the members of the Cypriot community. B. Séguin et al., 'Human Genomic Variation Initiatives in Emerging Economies and Developing Countries', *supra* Introduction, note 30; P. Marshall and B. Koenig, 'Accounting for Culture in a Globalized Bioethics' (Summer 2004) 32:2 *The Journal of Law, Medicine and Ethics* 252.

96 A. Kuper, 'Debate: Global Poverty Relief, More than Charity: Cosmopolitan Alternative to the "Singer Solution" ' (2002) 16 (1) *Ethics and International Affairs* 107.

97 This can be observed among the states participating in international trade negotiations under the auspices of the WTO: J.M. Finger and P. Schuler, *Implementation of Uruguay Round Commitments: The development challenge*, 1999, World Bank, Geneva.

economically advantageous for them, such powerful actors of the global economic order may recognise a corrupt or coercive government as the legitimate authority over a territory, or may do nothing while weak and powerless governments favour foreign interests over those of their own people.[98] Thus, the more powerful actors dictate terms that appear inescapable to those who are incapable of changing them. This conduct clearly influences how benefits and burdens are distributed globally and domestically. The growing global interconnectedness requires a critique of the fairness of the present global structure. If the global environment is one of justice, as argued here, domestic justice cannot take priority over global justice; both must be established in tandem.[99]

Applying lower moral standards to the basic global order involves arbitrary discrimination in favour of wealthy societies and against the global poor.[100] In fact, such global oppression by certain countries on the rest of the world does not meet minimum domestic standards of justice. Thus, rich and powerful countries impose a global economic order under which millions die each year due to poverty, and in which the gap between the rich and poor continually increases.[101] Since this would not be acceptable domestically, it therefore gives rise to a double standard for which there is no reasonable justification.

1.2.2.3 The place and role of states in the application of a theoretical framework of global distributive justice

Who owes justice to whomever it is justice is owed? Notwithstanding our shift in emphasis with respect to the role of states in evaluating justice, the rationale for envisioning distributive justice globally still allows states the possibility of remaining vital actors in such distribution. One of the crucial elements of the argument presented here is that equality between individuals should be considered the moral standard that would justify equitable access to

---

98 For example, many of the provisions of TRIPs, a document that has to be ratified by every member state of the World Trade Organization, reflect the views and demands of countries with powerful industrial lobbies for high levels of intellectual property protection. We will be coming back to this specific issue later in chapter 3 of the book. For a critique of this reality, refer to: T.W. Pogge, 'Cosmopolitanism and Sovereignty' *supra* note 18, at 108.

99 H. Shue, 'The Burden of Justice' (1983) 80 *Journal of Philosophy* 600, at 603.

100 Needless to say, the location of the world's valuable natural, wealth-creating resources is a matter of sheer luck. It therefore seems unfair that some people, because of a different control over resources, would prosper more and be less subject to discrimination than others. T. W. Pogge, *supra* note 70; C. Jones, *supra* note 10, at p. 73.

101 T.W. Pogge, *supra* note 70.

genetic innovations when required. Can states have a role to play in this framework? If so, what can it be within our cosmopolitan framework of global distributive justice?

To understand cosmopolitanism, the difference between moral and political structures must be clearly delineated. As demonstrated above, the main objective of our moral conception of justice is to attain some form of equality with regard to normal functioning and opportunity of every individual affected by institutional distributive arrangements. This does not mean that justice demands a total abrogation of sovereignty or the elimination of states. It instead requires an interpretation of sovereignty that does not constitute an arbitrary limit to the scope of justice.[102] Cosmopolitanism does not propose a best institutional structure for doing international politics. Indeed, principles of justice can continue to impose obligations for the satisfaction of individuals' rights on states and other institutional actors that may also be viewed as agents of international justice obligations.[103] Cosmopolitan justice does not rule out the importance of communities, as they may cultivate special bonds of sentiment, identity, and obligation among individuals. Thus, the fact that loyalties and connections are often associated with membership in a community is very important for some individuals and may still be taken into account under a cosmopolitan view of distributive justice. Special internal responsibilities can be significant if they do not take our attention away from the people who stand outside of the special relationship.[104] In this sense, a state-based world order may better serve human interests, with a perspective that includes everyone and can include special responsibilities for others' needs, insofar as the global theoretical basis of justice is acknowledged and respected.

In attempts to draw connections between the moral aspects of cosmopolitanism and the political reality in which states are major actors, different propositions for more porous boundaries have been promulgated. For example, Charles Jones proposes *qualified sovereigntism*, in which states and organisations of states maintain elements of their authority and sovereignty, but higher powers may supersede such authority if it does not meet the requirements of cosmopolitanism.[105] Wilfried Hinsch, by contrast, presents an international order with more or less independent states, united by contracts and

---

[102] O. O'Neill, 'Justice and Boundaries', *supra* note 58 at p. 69.

[103] C.R. Beitz, *Political Theory and International Relations*, *supra* note 1 at p. 271.

[104] C.R. Beitz, 'Cosmopolitan Liberalism and the States System', *supra* note 37, at pp. 130–131; C Jones, *supra* note 10; S. Scheffler, 'Liberalism, Nationalism, and Egalitarianism' in R. McKim and J. McMahan (eds), *The Morality of Nationalism* (New York: Oxford University Press, 1997).

[105] C. Jones, *supra* note 10, chapter 8, at 225 et seq.

mutual agreements that are governed by principles of global distributive justice and which apply directly to citizens of the world rather than to states.[106] As mentioned above, another proposition comes from Thomas Pogge, who suggests a form of vertical sovereignty in which governmental functions and areas of competence would be reallocated by their division and distribution to various coordinated political units. A form of centralisation would thus be established, but this structure could, at the same time, lead to the scattering of political allegiances and loyalties over both the new and the traditional groupings such as neighbourhoods, towns, counties, provinces, states, regions, and the world at large. [107]

There does not seem to be any contradiction in holding both that the ultimate moral focus for the analysis of international justice should be the interests of individuals and that justice development may take place within the basic structure of a decentralised order like the state system rather than a world government. As suggested by Andrew Kupler, normative cosmopolitan principles should consider 'individuals to be the normative epicentre of a system of functionally plural sovereignty'.[108]

In our specific context of analysis, states remain very important actors. Indeed, in the field of genetics, once products and services are available, further complex diagnostic testing steps are often required to identify needs, genes, susceptibility to diseases, and to target treatment. This type of procedure is likely to be undertaken at a population level where states' health care representatives, in their role as the existing present authority, can be important players in the distribution and allocation of genetic benefits to individuals in need. Moreover, the significant involvement of the private sector in genetics, especially from multinational pharmaceutical and biotechnology companies, calls for effective national intervention. In fact, the possibilities, in certain cases, for states to impose some sort of control over corporate conduct (for example, through price fixing, technology transfer, and corporate taxation policies) could have dramatic effects on the international distribution of the gains arising from genetic research. As such, states remain significant for our theoretical framework. Indeed, they would, at least in the actual world order, provide the agency required to perform obligations of international justice, whereas individuals, not just citizens, would represent the standard for consideration in determining how states ought to act.

---

[106] Hinsh, 'Global Distributive Justice' (2001) 32 *Metaphilosophy* 71.

[107] T.W. Pogge, 'Cosmopolitanism and Sovereignty', *supra* note 18, at pp. 99–108.

[108] A. Kupler, *supra* note 5; A. Kupler, *supra* note 17.

CONCLUSION

In this first chapter, I have made the argument for considering the global environment as the context of application for our framework of distributive justice. I adopted a cosmopolitan perspective that justifies the focus on individual human beings as the ultimate determinative basis, a standpoint particularly relevant for dealing with my specific issues of concern. Analysing some critiques of the global perspective on justice and refuting them with evidence on the limits of strict sovereignty and boundaries gave the perspective needed to determine how institutions involved in the distribution of the benefits of genetic research should operate and on what scale distribution should be handled.

The second chapter presents a specific argument for global distribution in health and genetics by setting up a scheme of health/health care justice at the global level.

2. An argument for global distribution in health

What are the reasons for intervention in health (in the process of accessing the good life)?

INTRODUCTION

The second chapter of this first theoretical part will elaborate an ideal conception of distributive justice in health to justify global access to genetics. I will establish normative grounds as the basis for my scheme of global health/health care justice, focusing on the special characteristics of health and on its crucial role in normal human functioning. This argument relies on the premise that health is a basic and essential good and that any reasonable account of justice must address the distribution of health care, resources, and services in the global order. I will then analyse the impact of normal functioning on the lives of individuals, using the criterion of the range of normal opportunities available to people. This will help to establish clear links between health problems, lack of access to the resources emerging from genetic research, and a diminution of the range of opportunities for which individuals of equal skill can build life plans. Then, I will extend the discussion to the global perspective on health as a way to propose some normative grounds for global distribution in health and genetic innovation. This discussion will highlight the specificity and universal importance of health. It will also flesh out my argument in favour of compensation for deviations from normal functioning and for the eradication of global health inequities over which we can have some form of control through distributive justice mechanisms.

2.1 CONCEPTION OF HEALTH JUSTICE

Some have argued that the pursuit of health should be embedded in social justice's broader quest to provide it with a stronger focus and a better understanding of the underlying social processes of health, and their fairness.[1] This

1 F. Peters and T. Evans, 'Ethical Dimensions of Health Equity', *supra* chapter

view is based on the argument that social inequalities in health essentially find their source in the basic structure of social, political, and economic institutions, as well as from consequent inequalities in poverty, income, and opportunity.[2] As demonstrated below, a theory of health justice remains linked in many ways to a more general all-purpose theory of justice. Nevertheless, health and, more specifically, genetics, raise unique issues that should be approached within a specific sphere of justice.

One very broad and idealised vision of health is 'a state of complete physical, mental, and social well-being, and not merely the absence of disease or infirmity'.[3] This definition is problematic in that it is so broad as to encompass many subjective conceptions of well-being and lead to a perfectionist conception of a healthy human being. Like Daniels and Borse, I define health as the absence of disease in a broad sense, which includes disabilities,[4] loss of abilities due to trauma and environmental harms, as well as other functional deficits.[5] In that sense, disease means any deviation from the normal functional organisation of a typical member of a species which comprises all impairment of functions.[6] Health therefore closely relates to *normal functioning*, a notion that will be analysed in greater detail in the next subsection.

Health is unlikely to ever be uniformly distributed among individuals. It can be influenced by various factors like individual biological variations, adequate nutrition, sanitary hygienic living and working conditions, chance, free and informed consent, and availability of preventive, curative, and rehabilitative medical resources and services.[7] Even if health and diseases are often thought to be beyond individual responsibility as the result of a natural lottery, namely environmental and socio-economic factors, some argue that the determinants of health over which we have control should be taken into

---

1, note 6 at 24–33; S.R. Benatar, A.S. Daar and P. Singer, 'Global Health Ethics: The Rationale for Mutual Caring' (2003) 79:1 *International Affairs* 107, at 122.

2 O. O'Neill, 'Justice Gender and International Boundaries' in: M.C. Nussbaum and A. Sen (eds), *Quality of Life* (Oxford: Clarendon, 1993) pp. 303, 315.

3 Preamble to the Constitution of the World Health Organization. Adopted by the International Health Conference held in New York, 19 June–22 July 1946, and signed on 22 July 1946. *Official Record of World Health Organization* 2, no. 100.

4 Which, in Doyal and Gough's terms, mean the 'consequent restriction or lack of ability to perform an activity in the manner or within the range considered normal for human beings' L. Doyal and I. Gough, *supra* chapter 1, note 42 at 172.

5 N. Daniels, *Just Health*, 2nd ed. (New York: Cambridge University Press, 2008) chapter 2, at p.12.

6 N. Daniels, 'Health Care Needs and Distributive Justice' (1981) 10 *Philosophy and Public Affairs* 146; N. Daniels, 'Equality of What: Welfare, Resources, or Capabilities?' (1990) 50 (supplement) *Philosophy and Phenomenological Research* 273, at 280.

7 N. Daniels, *Justice and Justification: Reflective Equilibrium in Theory and Practice* (Cambridge: Cambridge University Press, 1996) chapter 9, at p. 179.

consideration for distribution.[8] This argument calls for the elimination of only random inequalities that are not subject to choice, and has been called the *level playing-field* ideal.[9] Advocates of such a vision, *luck egalitarians*, consider responsibility to be at the centre of moral concerns, arguing that distribution should apply only to things over which individuals lack control and that are unrelated to free and informed choices. This position is controversial and does not always present the full picture. Indeed, ostensibly voluntary health-related behaviours are often driven by other socio-economic factors over which people frequently lack control. In fact, studies show that detrimental health habits do not necessarily always arise from free and informed will, but are often foreseeable results of poverty in childhood and beyond.[10]

This is why, even if personal responsibility for health and genetic characteristics can, in certain cases, be an interesting and complex argument, I will leave it out of our moral focus for the purposes of this discussion. Indeed, I argue for health equity as a way to achieve equality in opportunities, regardless of the cause of disease and how responsibly ill individuals can handle available opportunities. My vision of the international and national orders is more compatible with a system of cooperation that guarantees, in certain crucial areas, a '*safety net through which even the imprudents are never forced to fall*'.[11] This secures a range of basic needs which allows individuals to function in a community and choose from available opportunities.

It has been suggested that health inequalities may be considered inequities when they are unavoidable, unnecessary, and unfair.[12] Social inequalities in health can be perceived as a responsive indicator of the fairness of the basic social order. 'Unavoidable' individual health variations rooted in biological differences could be seen as acceptable if they were indiscriminately spread

---

8 D. Wikler, 'Personal and Social Responsibility for Health' (2002) 16:2 *Ethics and International Affairs* 47.

9 J. Roemer, *Equality of Opportunity* (Cambridge: Cambridge University Press, 1998); R. Arneson, 'Equality and Equality of Opportunity for Welfare' in L. Pojman and R. Westmoreland (eds), *Equality: Selected Readings* (New York: Oxford University Press, 1997); R. Arneson, 'Equality of Opportunity' *The Stanford Encyclopedia of Philosophy (Winter 2002 Edition)*, E.N. Zalta (ed), online: http://plato.stanford.edu/archives/win2002/entries/equal-opportunity (accessed: 26 January 2009).

10 J.W. Lynch, G.A. Caplan and J.T. Salonen, 'Why Do Poor People Behave Poorly? Variation of Adult Health Behaviours and Characteristics by Stages of the Socioeconomic Life course' (1997) 44:6 *Social Science and Medicine* 809.

11 E. Anderson, 'What is the Point of Equality?' (Jan. 1999) 109:2 *Ethics* 287, at 325.

12 G. Dahlgren and M. Whitehead, *Policies and Strategies to Promote Social Equality in Health* (Stockholm: Institute of Future Studies, 1991).

between social and geographical groupings and had nothing to do with education, income, or economic factors.[13] However, this is clearly not the case. Wealth and health disparities are constantly expanding within and between nations, both in terms of access to health care for individuals and of development and availability of adequate and population-specific treatments.[14] Health differences among individuals within different populations of the world are great and are often related to socio-economic factors such as income and education levels, the gap between rich and poor, public health measures, and access to health care and technologies. Therefore, in the field of genetics, the limits of what is unavoidable are unclear. Enduring biological/genetic characteristics can be directly caused by socio-economic factors. In fact, it would be unfair to acknowledge that individual members of a population are affected with unavoidable biological differences and diseases when the technology to prevent or cure those genetic predispositions and conditions exists, but is neither available nor affordable. Since health is not a good that can be directly transferred or allocated by distributive measures, our normative framework will focus on the factors that have a direct influence on individual health inequalities.[15]

In the Introduction to this book, I said a few words on how human genetic research is a very promising field for improving health globally. Many new products, including vaccines and drugs for common diseases, may eventually be based on genetic research. The human genome undeniably offers exceptional opportunities for understanding mechanisms of disease and developing new drugs and vaccines. Indeed, when we look at how genomics and genome-related biotechnology have developed in the past few years, we can say that advances in this field will likely, if applied correctly,[16] instigate important changes in the field of medicine and health care in the near future.[17] However,

---

13 F. Peter and T. Evans, 'Ethical Dimensions of Health Equity', *supra* chapter 1, note 6, at 28.

14 J. Sach, *The End of Poverty: Economic Possibilities for Our Time* (New York: Penguin Press, 2005).

15 D.M. Hausman, Y. Asada and T. Hedemann, 'Health Inequalities and Why They Matter' (2002) 10 *Health Care Analysis* 177; S.R. Benatar, A.S. Daar and P. Singer, supra note 1.

16 Many fear that genetics could also give rise to rampant abuses, like eugenics, and undesirable perspectives, like genetic determinism. Because of the many possible excesses and serious consequences they can have, a strict normative framework is needed to regulate genetic research and applications. These fascinating issues are not addressed in this book as we are starting from the premise that the benefits arising from genetics and relevant for global health will be the result of good quality scientific research undertaken and applied in strict compliance with appropriate legal and ethical principles.

17 As we have seen earlier – see Introduction – some, on the other hand, consider

new technology and health care innovations are simply beyond the financial reach of much of the world's population and could end up benefiting only a privileged minority, thereby increasing inequities in global health, as has been the experience thus far with certain drugs and vaccines.

This brings me to say that genetics is one field in which it has become vital to develop measures that will acknowledge the inequalities existing between individuals living in different countries of the world in terms of access and availability of the technology, undertake efforts to reduce them, and build safeguards for investment and research. To justify such distributive actions, I argue for a form of egalitarianism that asks for equal treatment of every individual in terms of bringing them to a level of normal functioning. In this framework, genetic benefits that can play a role in normal functioning should therefore be developed and made available to every individual in need, in line with a cosmopolitan vision insisting on equal consideration of individuals in the pursuit of global health.

2.1.1 Promoting Health to Avoid Serious Harm and Allow Normal Functioning

The elaboration of this framework for global distribution in health requires an examination of the rationale for the special treatment of health. In his *Discourses on Method*, Descartes wrote that the preservation of health was no doubt the chief of all goods and the foundation of all other goods of life.[18] The Millennium Development Goals (MDGs), adopted at the Millennium Summit of the United Nations in September 2000, call for dramatic improvements in the health of the poor. Indeed, health is central to the MDGs: 3 of the 8 objectives, 8 of the 18 targets, and 18 of the 48 indicators are health-related.[19] Furthermore, in a global consultation undertaken by the UN Secretary General Kofi Annan in 2001,

---

that this is an overstatement and that excess of optimism towards genetics should be moderated. For my part, I decided to focus on the positive impact that genetics is likely to have on global health in the future. For arguments in favour of this position, see: P.A. Singer and A.S. Daar, 'Harnessing Genomics and Biotechnology to Improve Global Health Equity' (5 October 2001) 294 *Science* 87. For arguments calling for moderate enthusiasm, see: T.M. Bubela and T. Caulfield, *supra* Introduction, note 24.

[18] R. Descartes, *Discourse on Method*, L. Lafleur (trans), (New York: Macmillan, 1960) at 85.

[19] UN General Assembly Resolution, *United Nations Millennium Declaration*, September 2000, A/RES/55/2, online: UN http://www.un.org/millennium/declaration/ares552e.pdf (accessed 26 January 2009); Executive Board Members, *Millennium Development Goals and Health Targets,* EB/Retreat/03/Discussion Paper 1, 11 November 2003, online: UN http://www.who.int/gb/ebmr/PDF/E/Millenium%20development%20goals%20and%20health%20targets.pdf (accessed 26 January 2009).

people around the world consistently indicated that good health was what they most desired.[20] Health's unique importance is acknowledged by many societies through special institutions and systems that ensure a more equitable distribution of health-related goods in comparison with other goods.[21]

Before saying more on the importance of health as a basic need, it is crucial to place it in a broader context where the avoidance of harm is the primary goal. In this context, the satisfaction of basic needs is an essential precondition to ensuring that serious harm is avoided. To qualify as serious harm, something has to prevent the pursuit of important individual goals. In Doyal's words, serious harm can be defined as 'the fundamental and sustained impairment of social participation and of basic needs for physical health and autonomy … which block new achievements which would otherwise have been real possibilities for the individual concerned'.[22] In this section, I argue that when serious health impairments can be avoided or prevented with existing products and services such as, for example, vaccines, drugs, and diagnostic tests, refusing or neglecting to provide them to individuals in need, wherever they are, is equivalent to inflicting serious harm. The superior basic human interest in health is shared by everyone, it represents an appropriate focus for a duty not to harm and ensures that individuals are in a position to profit from equality of opportunities, plan for a good life, and pursue their goals

Health needs sometimes vary between countries and geographical regions of the world because of environmental, nutritional, housing, and other socio-economic factors (as opposed to natural misfortune) and because the perception of health and illness can differ from one culture to another. In that way, therapeutic and preventive measures necessary to meet health needs can vary.[23] However, health as a basic human need can nevertheless be viewed as having a unique universal and objective significance.[24] In fact, as mentioned above, humans share similar genetic makeup and are physiologically alike although minor genetic variations can also entail important differences

---

20 Commission on Macroeconomics and Health, *Macroeconomics and Health: Investing in Health for Economic Development*, Dec. 2001, at 21, online on the WHO website: http://www3.who.int/whosis/cmh/cmh_report/report.cfm?path=cmh,cmh_reportandlanguage=english (accessed 26 January 2009); M. Johri and C. Barry, 'Health and Global Justice' (2002) 16:2 *Health and International Affairs* 33.

21 N. Daniels, *supra* note 5.

22 L. Doyal and I. Gough, *supra* chapter 1, note 42, at pp. 50–51.

23 K. Lederer, *Human Needs* (Cambridge: Oelgeschlager, Gunn and Hain, 1980).

24 That is why health is enshrined as a basic human right in international law, for example in the Universal Declaration of Human Rights, 1948, art. 25 and in the International Covenant on Economic, Social and Cultural Rights, 16 December 1966, art. 12.

between individuals which may require specific medical attention. This makes us all vulnerable to disease and health problems to different degrees, but because of the common characteristics we share with other human beings, we can say that 'universal and objective human needs do exist'.[25] In others words, even if each individual's unique genetic makeup might influence his precise health condition and his response to treatment in certain cases, in most cases, existing preventive and therapeutic methods will have similar effects on sick individuals and different degrees of health and disease will have comparable significance for almost everyone, anywhere.[26] As explained by Acharya, '[e]mpirically we are not likely to observe dramatic differences in the assessment of health in terms of what can be achieved with particular physical conditions across cultures'.[27] The objectivity of health-needs satisfiers can be demonstrated in genetic applications, with reliable diagnostic tests, for example; they could prove to be universally appropriate to test for the presence of a specific disease gene in susceptible populations anywhere. Another example of universality in health relates to physical pain, which, at a certain level, is objectively considered by almost everyone as something negative to be avoided.

Therefore, there seems to be a strong moral objection to the prevalence of different standards for health between different populations, particularly when these differences arise from socio-economic factors. As a result, some argue for universal agreement on some principles of justice and specific egalitarianism in the distribution of basic and universally important goods and services like those related to health. Basic needs like health care, food, water and shelter, all directly related to health, should therefore be regulated by special standards of egalitarian justice, different from those used to regulate the distribution of income and wealth.[28] Indeed, following this position, inequalities in health

---

25 L. Doyal and I. Gough, *supra* chaper 1, note 42, at pp. 2.

26 There are some exceptions to this, as if we take, for example, some physically impaired people who learn to develop amazing artistic skills and sensitivity, and for whom being ill probably does not have the same meaning and negative effects that it would have for a majority of people.

27 A.K. Acharya, *supra* Introduction, note 6, at 74.

28 It is important to mention here that since we are establishing an ideal theory of justice, we can consider all of those needs as universally basic and crucial. However, when we transpose this at the intermediate level of the non-ideal world, the lack of resources requires rationing the different basic needs like health, subsistence, education, etc. Fried clearly explains this dilemma and argues that focusing only on health care needs-satisfaction would prevent societies from pursuing other crucial social goals in C. Fried, *Rights and Wrongs*, (Cambridge: Harvard University Press, 1978) chapter 5; see also A.J. Culyer and A. Wagstaff, 'Equity and Equality in Health and Health Care' (1993) 12 *Journal of Health Economics* 431.

should be repaired without taking individuals' ability to pay for health care and services into consideration.[29]

But how and why is health a special good with an intrinsic value? Health is viewed as essential because it has a direct affect on every individual's normal functioning. 'Normal functioning' is a very broad concept and it is somewhat challenging to try to establish precise criteria for it. We cannot use this notion as a basis for global distribution if we consider it a subjective concept, adjustable to individuals' personal circumstances and variable depending on each individual's perception. For the purposes of this framework, we need a narrow standard of normal functioning, abstracted from personal choices and preferences. Aiming to define normal human functioning, Anderson says that:

> [t]o be capable of functioning as a human being requires effective access to the means of sustaining one's biological existence, food, shelter, clothing, medical care and access to the basic conditions of human agency, knowledge of one's circumstances and options, the ability to deliberate about means and ends, the psychological conditions of autonomy including the self confidence to think and judge for oneself, freedom of thought and movement.[30]

As it appears from this quote and as clearly explained by Daniels in his latest book[31] a thorough analysis of the concept of normal functioning requires that we look through a broader lens, not just referring to biomedical determinants but also interpreting it in the context of the different social, environmental, and genetic determinants of health distribution. However, for the needs of this discussion, I focus primarily on a form of egalitarianism that links biomedical and genetic resources with human beings' normal functioning to justify global distributive justice principles.

Being totally objective in defining normal functioning appears very difficult. In fact, as clearly explained by Boorse, values inevitably get involved when deciding which diseases impede normal function the most and therefore require more care and resources. There are many criteria that can be used to assess the priority of health care measures in pursuit of normal functioning as,

---

29 However, again, in front of the limited resources available in the non-ideal world, rationing within the broad field of health is also unavoidable. Although it is not the focus of this discussion, we will come back to how distribution of genetics' benefits could be undertaken to respect, as much as possible, our ideals of justice under section 2.2 of this chapter. J. Tobin, 'On Limiting the Domain of Inequality' (1970) 13:2 *The Journal of Law and Economics* 263; on rationing criteria in health: R. Plant and N. Barry, *Citizenship and Rights in Thatcher's Britain: Two Views* (London: IEA Health and Welfare Unit, 1992) at pp. 27–30; T.W. Pogge, *Realizing Rawls* (Ithaca: Cornell University Press, 1989) at pp. 178–188.

30 E. Anderson, *supra* note 11, at 317.

31 N. Daniels, *supra*, note 5.

for example, a treatment's potential for pain alleviation or death postponement. Our *objective* measure of individuals' normal functioning is directly related to the decent range of opportunities that are actually available to them. I argue that individuals should have access to the genetic products and services they objectively need to get to a level of normal functioning, which will allow them to take advantage of a decent range of opportunity (a concept to be analysed further in the next subsection).

Hence, health directly contributes to a person's basic ability to function in society, to interact in meaningful ways with other agents, and to live a life in which the pursuit of significant objectives and projects can occupy an important place.[32] Good health enables people to become educated, work, be productive, earn a salary, pursue personal and familial goals, and gain a certain degree of economic security, when possible in a given economic context. As indicated by Amartya Sen, health and education are constituents of development and are among the basic capabilities that give value to human life.[33] As it is not rare to see life plans compromised or considerably reduced by disease and poor health, good health is considered essential to a good life, one that is somewhat adequate in length and activity.[34] Good population health is also very important for poverty reduction and sustainable economic development, as societies affected by serious disease also experience considerable economic struggles and obstacles.[35] Therefore, since health can be characterised as a condition for normal species functioning,[36] it can be considered an objective need rather than a subjective preference or desire.

With respect to distributive justice, therefore, an *objective* criterion for health requires a standard independent from an individual's own evaluation, which might otherwise be representative of desires and preferences.[37] As such, if we consider that health is an objective need, vital to normal functioning, we could argue that many existing and future genetic technologies are equally necessary. Indeed, if genetics lives up to its great potential and proves useful in overcoming many serious and life-threatening health problems, those without

---

[32] S. Anand, 'The Concern for Equity in Health' (2002) 12:1 Harvard Center for Population and Development Studies Working Paper Series.

[33] A. Sen, *Development as Freedom* (New York: Knopf, 1999) Introduction and chapter 1.

[34] A.K. Acharya, *supra* Introduction, note 6, at 73.

[35] Commission on Macroeconomics and Health, *supra* note 20 at 22 and table 5 at 23.

[36] D. Braybrooke, 'Let Needs Diminish that Preferences May Prosper', in *Studies in Moral Philosophy*, American Philosophy Quarterly Monograph Series, No.1 (Oxford: Blackwells, 1968) at p. 90.

[37] N. Daniels, *supra* note 7, chapter 9 (health-care needs and distributive justice) at 179.

access to genetic services and technologies might be denied health benefits crucial for normal functioning. In fact, without access to genetic testing and screening technologies, individuals in deprived populations could lack the ability to prevent, cure, or reduce the severity of their conditions. Therefore, the more scientific developments in genetics lead to genetic testing and therapies for global health improvement,[38] the more disparities between the health condition of individuals with and without access to such technology will increase.[39] This will also likely create differences in the range of opportunities effectively available to individuals from those different groups.

2.1.2 Avoidance of Harm and Normal Functioning: Crucial Aspects of Equality of Opportunities

Having established that health is a central aspect of normal functioning, the argument now shifts to the role of wellness and normal functioning in the lives of individuals as a basis for distributive justice in health. The *equality of fair opportunity* has been popularised in Rawls' theory of justice as one of the main three principles, specifically that of 'justice as fairness'.[40] In short, it requires that persons with similar skills, talents, and ambitions have equal access to equivalent professional positions.[41] This concept does not require that opportunities be equal for all persons. In fact, unequal talents and skills among individuals are supposed to be covered by the application of the difference principle (inequalities have to be most advantageous to the least well-off). Rawls' theory of justice assumes a completely healthy population. It is therefore not designed to deal specifically with issues of health inequalities and health care distribution, and sets aside individual differences resulting from

---

38 As it is planned in this report, which presents 4 top genetic biotechnologies that could improve health in developing countries: Program in Applied Ethics and Biotechnology and Canadian Program on Genomics and Global Health (University of Toronto Joint Center for Bioethics) *supra* chapter 1, note 2.

39 M.J. Mehlman and J.R. Botkin, *Access to the Genome: The Challenge to Equality* (Washington, DC: Georgetown University Press, 1998).

40 Indeed, Rawls argues that rational individuals who would not know their position in a society would desire, after ensuring equal distribution of basic liberties, that two basic principles govern distribution in their institutions: equality of opportunity (professional) and the difference principle (that all inequalities in the remaining social goods be to the greatest advantage of the least well-off). J. Rawls, *supra* chapter 1, note 11; J. Rawls, *supra* Introduction, note 49.

41 N. Daniels, B. Kennedy and I. Kawachi, 'Justice is Good for Our Health: How Greater Economic Equality would Promote Public Health' (February/March 2000) 25 *Boston Review*, online: Boston Review http://www.bostonreview.net/BR25.1/daniels.html (accessed 26 January 2009).

disease.[42] Rawls only requires a fair distribution of basic liberties, opportunities, and economic resources. One's health, positive freedom, and actual capacity to convert such factors into normal functioning and well-being (professional advantages, well-being, wealth and so on) do not figure directly into his view.[43] It can be understood as setting up the justice framework for regulating distribution of key health determinants that are subject to social control, such as the structure of social organisation, government policies, wealth distribution, income inequality, poverty, and so on. Some also argue for the application of Rawls' difference principle to health concerns, which would require granting priority to the least-favoured groups in society with respect to health matters and to improve the health of the poorest in society in order to justify health inequalities.[44]

However, the pursuit of the kind of distributive justice promulgated here requires a system of distribution that meets health needs fairly at the 'point of delivery'.[45] This vision is that of Norman Daniels, who extends the theory and notion of equality of opportunity to cover health care and adapt it to the reality of disease and disability.[46] He believes that health care should be something to which we should have equal access in order to improve our health status and attain a level of normal functioning that allows access to a normal range of opportunities in other spheres of activities; or, in other words, to become efficient 'converters of primary goods'.[47] For Daniels, opportunity not only refers to the professional area but should also be considered as the portion of an individual's autonomy and liberty available for the achievement of various goals and undertakings, plans of life, and conceptions of the good.[48] In fact, when we argue for the moral equality of individuals, it means that we believe that everyone should have equal prospects to plan for a good life, should be entitled to participate and take part in their community's life and develop their individuality,[49] and that disparities in possibilities over which we

---

42 Ibid.

43 He has been criticised on this point by Amartya Sen, who argues that Rawls' focus obscures the difference between sick and healthy individuals in terms of capability of functioning. For more details refer to: A. Sen, *The Standard of Living, The Tanner Lectures* (Cambridge: Cambridge University Press, 1987); A. Sen, *Commodities and Capabilities* (New York: Elvier Science Pub., 1985).

44 S. Marchand, Daniel Wilker and B. Landesman, 'Class, Health and Justice' (1998) 76:3 Milbank Quarterly 449.

45 N. Daniels, B. Kennedy and I. Kawachi, *supra* note 41.

46 N. Daniels, *Just Health Care*, *supra* note 6 at pp. 36–48.

47 N. Daniels, *supra* note 6, at 279.

48 N. Daniels, *supra* note 6; see also A. Sen, *The Standard of Living*, *supra* note 43.

49 L. Doyal and I. Gough, *supra* chapter 1, note 42, at p. 91.

have control should be eradicated. Many external factors such as lack of water and food, severe poverty, political and socio-economic instability, and lack of access to health care goods and services can influence opportunities. Taking the example of health, if people with equal skills, talents, and ambitions are entitled to fair, equal opportunities and yet different degrees of access to health care technologies and resources are tolerated, some individuals will accordingly lack the same advantages with respect to those opportunities, since disease can considerably affect one's capacity to take advantage of available opportunities.[50]

Even if health problems will not always impact negatively on individuals' goals, projects, and level of happiness, it appears that individuals must possess a certain degree of physical health to participate in life, and accomplish projects in a cultural, personal, or professional context. They therefore ought to have an equal chance to obtain health care/genetic technologies and services to attain a level of objective normal functioning that will then allow them to profit from a decent range of opportunities, given their skills and talents. Indeed, even when societies establish measures to help and include people with disabilities, serious health impairments can still mean major limits on the range of opportunities that would otherwise have been available to someone with particular talents and skills.[51] Daniels' view follows from Rawls' theory of justice but does not depend upon it. Instead, Daniels adopts a notion of opportunity that is far broader than Rawls'. Rawls is interested only in access to professional positions and careers. Daniels goes one step further and associates the right to health care resources with the personal and social factors required for a good life.[52]

In the context of this analysis, the expression *health care* includes genetic technologies, tests, drugs, vaccines, and services available for preventing diseases and providing early diagnosis and treatments to improve the health status of individuals affected with specific diseases. Health care can have many functions for sick individuals, such as improving life expectancy, decreasing pain and suffering associated with their conditions, and providing tools for the prevention and treatment of diseases. In other words, 'it maintains, restores or compensates for the loss of functioning that is normal for a member of our species'.[53] Therefore, health care should be distributed on a

---

50 N. Daniels, *supra* note 6.

51 A. Buchanan et al., *From Chance to Choice: Genetics and Justice* (Cambridge and New York: Cambridge University Press, 2000) chapter 7.

52 Ibid.; L.M. Fleck, 'Just Health Care (II): Is Equality Too Much?' (1989) 10 *Theoretical Medicine* 301.

53 N. Daniels, 'The Genome Project: Individual Differences and Just Health Care' in: T.F. Murphy and Marc A. Lappé (eds), *Justice and the Human Genome Project* (California: University of California Press, 1994) chapter 7, at 110.

more equal basis rather than according to an individual's motivation and ability to pay.

Unlike preferences, tastes or desires, health relates to objective needs, the fulfilment of which allow an individual to access a normal opportunity range, to build up life plans for which he is suited, to establish relationships with others, and to develop different interests. As such, the crucial difference between treatment and enhancement in genetics is important. In referring to equality of opportunity in our framework of justice, it is not intended that genetic research should bring everyone to the same wellness and happiness level since medicine does not exist to make everyone happy and equal in terms of their skills and talents. Instead, it means equality in terms of access to genetic services and technologies that can be used for treatment of those who have an *objective medical need*, or in other words access to what is needed to attain *normal functioning* as an objective measure of health. This way, genetic technologies and services can be employed to bring individuals as close to the *normality level* (being healthy) as possible by removing obstacles in the way of access to a normal opportunity range.[54] One example of genetic intervention that has been proposed to ensure fair access to a normal opportunity range is germ-line and somatic genetic engineering in embryos to correct serious genetic defects before birth.[55] Notwithstanding the numerous ethical and scientific problems that could arise with this technology,[56] it may have considerable potential for the protection of fair equality of opportunities in our global structure if made available to everyone in cases where there is absolutely no doubt that the defective genetic makeup would lead to serious disease.

To conclude this section, in order to support the foundation for a specific theoretical framework for global distributive justice in health and, more specifically, in genetics, it has been established that health, as a vital element for normal functioning, greatly influences the range of normal opportunities available to individuals. Indeed, the impairment of normal species functioning reduces the range of opportunity in which we may construct our life plans. Therefore, the needs associated with normal species functioning can be qualified as objectively important since they correspond to the great interest people have in maintaining a normal range of opportunities. Attaining normal functioning to be in a position to benefit from equality of opportunity is thus crucial and is supposed to be protected against interferences from persisting

---

[54] N. Daniels, *supra* note 6.

[55] M. Leonard, *supra* chapter 1, note 4 at 133.

[56] Arising problems could, for example, be related to the fact that the technology is not yet at all safe and usable, that it would be extremely expensive (read impossible) to make it available to everyone, that it could lead to eugenic applications, etc.

inequalities. Having presented the reasons why health warrants special treatment, I now need to expand my discussion to the global aspect of health.

2.1.3 Global Perspective on Health

The argument presented above in regards to equality of opportunity and its application to health had originally been developed to apply only to citizens of a given society. As such, it depended on the idea that opportunities available to individuals are unique to a given society since they depend on specific characteristics of that society, such as its level of material wealth, its economic and technological development, its cultural particularities, its conception and shared understanding of justice, its administrative and institutional structures and regulations, and so on.[57] Therefore, it is argued that the same disease in two different societies will likely reduce opportunities in different ways, with the result that their significance would be differently assessed.

However, to support the argument that equality of opportunity should prevail within states but not at a global level would be to attribute certain rights to members of a society while denying them to others who are not part of a national system of cooperation. As discussed at length in the previous chapter, I do not agree with such a view that prioritises a domestic conception of distributive justice where states are the primary agents of justice and where members of a political community are entitled to special rights by virtue only of their membership in this society.[58] This argument provides a good basis for domestic equality of opportunities, but how does it justify rejecting a similar concept at the global level? Those in favour of a domestic application of this principle to health often argue that the major financial contribution of societies' members to support their health system (programmes, services, therapeutic products and so on) justifies limiting equality of opportunity to the national level. However, potential holders of a right to equality of opportunities within a society have not necessarily begun to participate in the system of mutual cooperation. In fact, like non-citizens, children and teenagers often did not engage in the economic system that allowed the creation of the goods and services that would be subject to equality of opportunities.[59] Moreover, if the enforcement of domestic equality of opportunity would oppose situations where people have fewer prospects due to their class or income, it would seem to follow that analogous situations in which people have fewer prospects according to their nationality, citizenship, or geographic location should also be opposed.[60]

---

57 N. Daniels, *supra* note 6; M. Walzer, *supra* chapter 1, note 11.

58 D. Harris, *supra* chapter 1, note 60, at pp. 56–57, 86, 103–104.

59 T.W. Pogge, 'Cosmopolitanism and Sovereignty', *supra* chapter 1, note 18.

60 S. Caney, *supra* chapter 1, note 60.

The cosmopolitan perspective of justice proposed in the first chapter is very relevant in the health sector, where different societies interact within a rich global structure of political and economic institutions that considerably affect health prospects for everyone. Therefore, the moral reasons to take responsibility for non-citizens' health problems caused by the global economic order are of similar importance when compared to co-citizens' health problems resulting from the domestic economic order.[61] This argument derives from the principle defended in the first chapter, that individuals, wherever they are, should be the most important standard of concern in establishing basic principles of justice. Therefore, it follows that we should not determine the significance of disease and lost opportunities on a society-based model. This application of the cosmopolitain perspective to global health and justice is one important contribution of the present book.

The normal opportunity range can vary between countries in terms of the nature of the actual opportunities, which include the types of careers available, the most rewarded physical and intellectual talents and skills, and the nature of possible undertakings and life plans in a given environment. Consequently, some, like Daniels, argue that since different societies call for different health care measures, the normal opportunity spectrum is relative to each society.[62] However, like the ability to function normally, the general types of opportunities to pursue life and career undertakings, to use some form of language and basic social ability, or to engage in some form of labour, should be the same for everyone regardless of their nation, state, or ethnic group.[63] The evaluation of the range of opportunities at the global level, as at the national level, may be generalised from an examination of the variety of possible prospects and from a subjective perspective regarding such opportunities. As Martha Nussbaum indicates, certain objectively important needs, like health, are valued by everyone since they relate to universally shared priorities. This can allow the creation of an international standard for evaluating opportunities on a global scale.[64] Consequently, the rationale for accepting the principle of equality of opportunity within the state also requires that we endorse a principle of global equality of opportunity, the whole of which contributes to the framework for global distributive justice. Indeed, the goal of global equality

---

61 Refer to Pogge's interesting discussion on this point: T.W. Pogge, *Justice in Health Care: Reflexions on the Foundations of Health Equity*, Working Paper Series, Cambridge, MA, Harvard Center for Population and Development Studies, 1999.

62 N. Daniels, *supra*, note 6.

63 H. Shue, *Basic Rights: Subsistence, Affluence and US Foreign Policy* (Princeton: Princeton University Press, 1996) at pp. 59–60, 159 et seq; L. Doyal and I. Gough, *supra* chapter 1, note 42, at p. 181.

64 M. Nussbaum, 'Human Functioning and Social Justice: in Defense of Aristotelian Essentialism' (2 May 1992) 20 *Political Theory* 202, at 216–223.

of opportunity in health is becoming increasingly essential, especially in light of the critical health problems affecting individuals who live in poorer countries where genetic technologies and services may not be adapted, available, or affordable. This leaves millions of individuals with reduced access to a normal range of opportunities in many spheres. This quote from Acharya summarises my position on the importance of the global aspect in equality of opportunities:

> It will be most likely agreed that children should be afforded the same chances for all possible future jobs, political offices, and opportunities. Nearly all children should be given an equal chance of survival conditional on their congenital status. A child should be considered to be especially disadvantaged if he or she, as an adult, will not be capable of qualifying for most types of employment in a given region when another child in a different region with a similar condition could obtain employment.[65]

However, it is important to mention that although we should ensure that, on the global scene, every individual of comparable skill and talent should be healthy enough to access similar sorts of opportunities, it does not mean that they should all be provided with identical medical attention and health care. Indeed, since uniformly broad categories of opportunities will actually give rise to different opportunities in different settings and environments, adapted health care standards will be required to bring individuals to a level of normal functioning in different societies. For example, if we consider the opportunity to pursue career undertakings or, even more generally, to engage in some form of labour, it appears that bringing people to a level of normal functioning will require different types of health care measures depending on the context of their respective societies. Indeed, a normal functioning standard in career and labour opportunities could mean something different in an industrialised context with established resources and infrastructure to treat people with disabilities than in a poorer, more rural/agricultural reality that lacks those same resources, and where the same diseases consequently lead to considerable burdens. Although the basic framework for distribution remains the same, genetic technologies could be harnessed in specific ways to respond to different countries' health needs and to bring individuals to the level of normal functioning they need in order to be able to take advantage of given opportunities and be functional in the environment in which they live.

---

65 A.K. Acharya, *supra* Introduction, note 6, at 74.

2.2 NORMATIVE GROUNDS TO OPERATE DISTRIBUTION AND PREMISES UPON WHICH TO CLAIM HEALTH EQUITY AND FAIRNESS

2.2.1 What Can Constrain Distribution in Health? What Kinds of Responsibilities do the Affluent of the World Have Towards the Global Disadvantaged?

The concept of global equality of opportunity as applied to health is an essential stand point for analysing the global distribution of health care, technologies, and services within a framework of justice. The role of the principles of distributive justice is to determine the fair distribution of genetic benefits that may be produced by global cooperation. Having established a rationale for fair distribution in this specific area, we will now address how it should be undertaken. To this end, various normative tools are available, such as right- and duty-based theories and the global application of Rawls' difference principle.[66] This discussion is important to resolve the ambiguity that prevails with regard to the moral grounds of any requirement to assure global health.[67]

2.2.2 Rights (To Equality in Opportunities)

According to Feinberg:

> Legal claim-rights are indispensably valuable possessions. A world without claim-rights, no matter how full of benevolence and devotion to duty, would suffer an immense moral impoverishment ... A world with claim-rights is one in which all persons, as actual or potential claimants, are dignified objects of respect, both in their own eyes and in the view of others.[68]

The purpose of this section is not to present a comprehensive analysis of the notion and desirability of rights, but to refer to a framework in which rights can be conceived as a potential basis for the requirements of our theory of global justice in health. I will return to rights discourse later in the book, when I analyse the legal framework and system of international socio-economic human rights.

The acknowledgement of rights is an important aspect of a theory of justice. It is, in fact, what often justifies the restrictions on action or the imposition of

---

66 C. Jones, *supra* chapter 1, note 10.
67 L.M. Fleck, *supra* chapter 1, note 6.
68 J. Feinberg, *Social Philosophy* (New Jersey: Prentice Hall, 1973) at pp. 58–59.

duties to act in certain ways. If we consider rights an important factor in a global distributive justice theory, they will accordingly impose restrictions on distributive arrangements (social, political, and economic) supported by the global order. In fact, we can consider that a just distribution is one in which each individual obtains what he or she can claim by right. When rights are established and recognised, they shift the burden of proof onto those who decide not to respect them. Indeed, as indicated by Will Kymlicka, justice can be considered the system of entitlements upon which people can base their demand for recognition of legitimate claims for resources and opportunities.[69]

The need to rectify injustice created by inequitable distributions can find its source in rights. Rights must be based on property or on something of crucial importance for its possessors.[70] They are often envisioned as a basis for justifying demands and imposing obligations. A right-holder can require that the content of his or her right be guaranteed. Rights are recognised and understood as being grounded in the basic interests that individuals have in the content of those rights. An argument for a right 'is an argument showing that an individual interest considered in itself is sufficiently important from a moral point of view to justify holding people to be under a duty to promote it'.[71] Some argue for the protection of basic rights associated with the primary necessities and preservation of human life. Such basic rights can emerge from needs shared by every human, like food, water, shelter, and health care.[72] Rights allow us to associate human well-being and related obligations. They can be ranked according to the nature of the interest they help to defend, as well as by their normative weight.[73] We will return to the related duties and obligations in the next part of this chapter.

As discussed above, health is critically important for individuals and represents a basic human interest. As we will see in more detail in the fourth chapter of this book, the right to health is controversial but is recognised as a human right in international law treaties[74] and requires positive action in terms

---

69 W. Kymlicka, *Liberalism, Community and Culture* (Oxford: Clarendon Press, 1989) at p. 234.

70 C. Jones, *supra* chapter 1, note 10

71 J. Waldron, *The Right to Private Property* (Oxford: Clarendon Press, 1988) at p. 3; on the interest notion see also: J. Waldron, *Liberal Rights: Collected Papers 1981–1991* (Cambridge: Cambridge University Press, 1993) at pp. 11 and 212; J. Raz, *The Morality of Freedom* (Oxford: Clarendon Press, 1986) at p. 166.

72 H. Shue, *Basic Rights, supra* note 63.

73 O. O'Neill, 'Hunger, Needs and Rights', *supra* chapter 1, note 44; P. Jones, *Rights* (London: MacMillan, 1994) at pp. 13–15.

74 For example, the right to health has been codified in: Universal Declaration of Human Rights, art. 25; International Covenant on Economic, Social and Cultural Rights (ICESCR), G.A. Res. 2200 (XXI/, UN GAOR, 21st Sess., Supp. No. 16, at 49,

of resources or actions. Interpreted broadly, the right to health can include the right to health care and the right to genetic technologies, especially if we consider that they could become the new standard of care needed to achieve acceptable standards of health and a broader and universal right to equality of opportunity. In fact, in the field of health, it is inadequate to refer only to a basic right to emergency health care and subsistence if access to adequate and adapted health care remains reserved for a privileged few.[75]

The very concept of rights is not accepted by everyone and is, in fact, often criticised.[76] One reason for this is that rights remain meaningless in cases where taking advantage of them is not a real option for the right holders.[77] This objection seems to apply more to civil and political rights. In fact, even if it appears easy to grant rights such as the right to vote, the right to freedom and to security, or the right to free speech, if personal health, material, and economic conditions of the right holders are such that they cannot take advantage of them, these rights lose much of their value. While some argue that the strong socio-economic component of many traditional civil and political rights justify their extensive protection,[78] it may make more sense to eliminate what is now a clear separation between civil and political rights and economic, social, and cultural rights.[79] In fact, since poverty and ill-health are considerable obstacles to a satisfying human existence, economic and social rights

---

UN Doc. A/6316 (1966) art.12; *Convention on the Elimination of all Forms of Discrimination against women*, G.A. res. 34/180, 34 U.N. GAOR Supp. (No. 46) at 193, U.N. Doc. A/34/46, entered into force 3 September 1981, art.12; *Convention on the Rights of the Child*, G.A. res. 44/25, annex, 44 U.N. GAOR Supp. (No. 49) at 167, U.N. Doc. A/44/49 (1989), *entered into force* 2 September 1990, art. 24; See also, *African Charter on Human and People's Rights*, adopted 27 June 1981, OAU Doc. CAB/LEG/67/3 rev. 5, 21 I.L.M. 58 (1982), *entered into force* 21 October, 1986 art. 16; *Additional Protocol to the American Convention on Human Rights in the Area of Economic, Social and Cultural Rights*, 1988, art. 10.1.

[75] S. Caney, *supra* chapter 1, note 60.

[76] Indeed, many scholars denounce the use of rights as a way to structure and regulate a society, arguing that it creates entitlements without envisioning them within the context created by the society's structures and institutions. For a comprehensive exposé on the criticism of the character and the desirability of rights refer to: M. Sandel, *Liberalism and the Limits of Justice* (New York: Cambridge University Press, 1998).

[77] In *A Theory of Justice*, Rawls recognises that the value of liberty is directly linked with the material conditions necessary to experience it.

[78] J. Waldron, *supra* note 71 at pp. 7–8; M. Jackman, 'Poor Rights: Using the Charter to Support Social Welfare Claims' (1993) 19 *Queen's Law Journal* 65.

[79] The strong connection between the two categories of rights and the fact that there is no hierarchy between them has been specifically emphasized by the UN: United Nations Committee on Economic, Social and Cultural and Economic Rights, *General Comment No. 3(1),* 2002, online: UN http://www.unhchr.ch/html/menu2/6/cescr.htm (accessed 30 May 2006).

should rank alongside civil and political rights,[80] and all of these rights should therefore be envisioned as required values for a better world.[81] This has a critical importance with respect to health issues since it seems essential to ensure that each individual has equal access to appropriate health care, technology, and resources in order to achieve a normal range of opportunities in other spheres and therefore be able to take advantage of other civil and political rights.

Another objection to rights discourse relates to the notion of solidarity, which holds that granting rights to people can have the effect of isolating them from each other.[82] In fact, proponents of this view argue that a focus on individual rights allows neglect of collective responsibilities – for example, in allowing some individuals to control the majority of wealth while many others, though possessing rights, end up with a lot less than they need. This creates social division that can prevent the establishment of a truly solid community. In response to this argument, others assert that any defensible account of rights demands a strong social framework that would facilitate mutual sharing of the benefits of rights and burdens of duties. For advocates of this view, rights are part of a 'reciprocal universality' and thus it is impossible to see right-holding as a totally selfish and individual experience.[83] As Bowles and Gintis maintain, 'the discourse of rights has served as a source of bonding and a framework for the expression of group demands, rather than reflecting a social philosophy or a political ideology'.[84] The cosmopolitan perspective holds that where the importance of vital interests of each individual should be acknowledged, regardless of his or her location, nationality, or citizenship, rights are essential to achieving the moral distributive ideal. In fact, when they are recognised, integrated within a society's functioning structure, and taken seriously, they can provide a basis for impartial and equal individual recognition, as well as for community solidarity and cohesion. As such, from the cosmopolitan perspective, rights cannot remain based on individual

---

80 B. Terence and R. Dagger, *Political Ideologies and the Democratic Ideal*, 5th edn (New York: Addison Wesley, 2004).

81 J. Waldron, *supra* note 71 at pp. 7–8; G.B. Herbert, *A Philosophical History of Rights* (London: Transaction Publishers, 2002).

82 J. Waldron, *Nonsense upon Stilts: Bentham, Burke and Marx on the Rights of Man* (London: Methuen, 1987); R. Vincent, *Human Rights and International Relations* (Cambridge: Cambridge University Press, 1986) part 1.

83 For example refer to: C. Jones, *supra* chapter 1, note 10, at pp. 80–83; A. Gerwirth, 'Rights', in L.C. Becker and C. Becker (eds), *Encyclopedia of Ethics* (London: Garland, 1992) at p. 1108.

84 S. Bowles and H Gintis, *Democracy and Capitalism: Property, Community, and the Contradictions of Modern Social Thought* (New York: Basic Books, 1986) at p. 170.

experiences, but must be integrated in a collective framework where the structure in place imposes global respect and recognition. This aspect will be treated later when I discuss the existing human rights framework.

Also, the enforcement aspect of rights gives rise to scepticism about the concept of a right to health care, genetic technologies, and resources. Critics emphasise the material obstacles to supporting and implementing welfare claims, that mainly relate to the scarcity of resources and the identification of accountable actors and institutions.[85] They maintain that rights should not only cover the essence of what should be granted, but should also talk about what would be necessary to achieve them and by whom they should be respected. By focusing on enforcement, such critiques emphasise the abstract aspect of rights. Onora O'Neill for example, argues that 'if the claimants of supposed "rights" to food or development cannot find where to lodge their claims, these are empty "manifesto rights" which would be equivalent to having no right at all'.[86] She argues that rights-talk is rhetorically powerful, but is not ethically founded because it does not deal with the powerful actors who could do something about international injustice.

However, with this type of argument about legal enforcement, the inherent normative importance of a focus on welfare rights is often ignored as it assumes a strict legalist *Hohfeldian* vision of rights. In the early 1920s, Hohfeld published a very influential text in the field of rights, *Fundamental Legal Conceptions*,[87] in which he approaches diverse theoretical differences informing rights discourse. According to Hohfeld, the concept of 'rights' gives rise to correlated duties and is thus best defined as *claim-rights* against another party who, as a consequence, becomes legally obligated.[88] However, this is not the only way to envision rights, especially not human welfare rights, the substance of which should not be subject to the same restrictive juridical conceptualisation.[89]

The types of rights on which we could establish our justice framework (rights to health and health care, to genetic technologies and resources, to opportunities) relate to human values that are essential to people's welfare. Rights can give rise to different types of duties, some positive (such as, for example, to perform actions, to assist) and others negative (such as to refrain from performing certain actions).[90] Therefore, a restrictive conceptualisation

---

85 R. Nozick, *supra* chapter 1, note 6, at pp. 112–114.

86 O. O'Neill, (2000), *supra* chapter 1, note 16, chapter 7 at p. 115 et seq.

87 W. Hohfeld, *Fundamental Legal Conceptions* (New Haven: Yale University Press, 1923 and Aldershot: Ashgate, 2001).

88 Ibid. at pp. 38 and 46.

89 G.B. Herbert, *supra* note 81, at p. 280.

90 J. Narveson, *The Libertarian Idea* (Philadelphia: Temple University Press,

of rights, with required correlative positive duties, is not necessarily always applicable to health, health care and technologies, or to the right to equality of opportunities in practice. Indeed, they can exist without the actual claim of specific positive duties against identifiable actors as they are associated with the justice values and principles to which we aspire globally. However, some argue that human rights, like the right to health, certainly give rise to negative duties not to harm, and these commentators refuse to endorse an institutional order that entails avoidable and foreseeable violation of those rights.[91] The rights to health/health care, technologies, and resources are inherently important to protecting individuals' welfare and vital interests; this alone justifies their recognition as normative grounds for justice in a just global structure. The fact that the related positive duties associated with these rights are sometimes contested (or not yet allocated to particular actors or institutions) does not mean that those rights do not exist and cannot be an important part of our ideal justice theory. As Gewirth puts it, '[i]t is not enough to say that rights-enforcement incurs costs; there is the prior question of what there is about rights that makes them worth the cost'.[92]

This being said, in order for the distributive justice theory in health care to have political significance, it needs to have an institutional component. Justified rights must be adequately protected by required duties and obligations to refrain from harm, defend the interests of rights-holders, and facilitate the enforcement of their rights against particular agents.[93]

2.2.3 Obligations/Duties (To Redress Distributive Injustice)

2.2.3.1 Who owes justice to whomever it is justice is owed?

The practical and institutional aspects of obligations require examination, particularly in connection with the concept of obligation and duties in relation to justice. We have seen that Daniels' theory allows us to consider health care access as directly related to the normal range of opportunities. Health care and genetic resources and services should thus be distributed in a way that allows normal functioning and consequently ensures equality of opportunities. This may give rise to a right to equality of opportunities and to correlative obligations.

1988) at 57; R. Cruft, 'Human Rights and Positive Duties' (Spring 2005) 19:1 *Ethics and International Affairs* 29, at 30.

91 T.W. Pogge, 'Severe Poverty as a Violation of Negative Duties' (Spring 2005) 19:1 *Ethics and international Affairs* 55, at 66–68.

92 A. Gewirth, 'Are All Rights Positive?' (2001) 30:3 *Philosophy and Public Affairs* 321, at 330.

93 J. Feinberg, *supra* note 68, at p. 59.

2.2.3.2 What duty/obligation?

Following Henry Shue's classification, obligations can arise from three different kinds of duties, which are either positive or negative: the negative duty of avoidance, the positive duty of protection and the positive duty of aid.[94] The global distributive justice theory would include several related duties such as avoidance of harm to global health, institutional protection against harm, provision of relief, provision of aid development, and redistribution of genetic resources. The aim, purpose and responsibility of the different actors involved will determine the nature of the duties that will be assigned to them. For example, as multinational corporations are not designed to protect human rights, but to make profits and gain power in their sphere of activity, they may merely have to act so as to avoid violating human rights (negative duty of avoidance) without necessarily having to promote human rights, as other actors might be otherwise compelled to do.

Health deprivation and aggravation may be caused by many actors because of a variety of factors. This is why it can be tempting to say that since health deficiencies are the responsibility of so many agents, they are not specifically anyone's responsibility. There are different arguments as to how we should allocate duties in health-related spheres. Shue's response is to propose prioritising rights' protection. He gives the fulfilment of basic rights priority over all other non-basic rights, which he, in turn, gives priority over satisfaction of preferences and cultural advancement.[95] If we apply this priority principle to the access to health, health care, technologies, services, and resources, duty-bearers could find themselves obligated to fulfil those rights before anything else. As discussed above, equitable access to good health is a prerequisite to equality of opportunities in many crucial spheres of life. It is critical to every individual's complete personal development; as such, the actual level of inequality in this area contributes to preserving a degrading and unfair level of inequality. Other important factors to consider in determining the scope of justice obligations are the urgency of the condition and the cost of delivering the required assistance. In other words, Shue makes special responsibilities for health depend on emergency and seriousness of interests without necessarily taking responsibility for health into account. This demands positive distributive duties even in cases where no responsibility can be attributed. This is particularly relevant for global health since the actual situation has become worse for so many years for so many different political, environmental, and socio-economic reasons – a context that makes it very hard to attribute direct responsibility to specific actors. Where the more fortunate can do something

94 H. Shue, *Basic Rights, supra* note 63, at p. 52.
95 Ibid., at p. 118.

about serious diseases and suffering without unduly burdensome costs, they should do so.[96]

Another way to envision duty allocation is to focus on responsibility for deprivation, giving rise to a duty to avoid causing harm to others. This can be considered in many different ways, one being that everyone who supports an unjust global structure is responsible for the injustice it creates,[97] and another being that a failure to secure universal and basic needs (health-related needs for example) implies failing in meeting one's duty to refrain from harm.[98] We can cause harm by exposing others to extreme poverty and health deficits, and failing to intervene when, for example, life-saving knowledge and products are available but inaccessible for structural reasons. In such cases, the negative duty to avoid harm can require positive obligations, that is, that we take positive steps to ensure that the legal structure we support is not encouraging these health gaps. Individuals can also be harmed if others fail to recognise the importance, objectivity, and universality of their crucial health needs, and consequently restrict them in the pursuit of their life goals, as discussed above. Again, the duty not to harm extends to agents and implies that individuals should participate in creating and supporting institutions that can help meet basic needs and relieve major suffering.[99] The ideal just global structure I argue for must adhere to global distributive mechanisms to avoid such harm.

In addition to the basic need and to the responsibility views, a third way to allocate duties for health improvement is Buchanan's natural duty of justice, which requires the creation of mechanisms to provide universal access to just institutions, even outside established schemes of cooperation.[100] This idea is based on a cosmopolitan vision that treats all human beings with equal consideration by respecting and protecting their most basic needs and helping to create a global context where individuals will have access to institutions that secure their fundamental rights. It is similar to the duty not to harm, but requires positive duties to establish institutions and legal principles to regulate

---

[96] H. Shue, *Basic Rights, supra* note 83 at pp. 17, 59–60, 159–161, 164–166, 168–180; T. Pogge, 'Eradicating Systematic Poverty: Brief for a Global Resources Dividend' (2001) 2 *Journal of Human Development* 59; T. W. Pogge, 'Responsibilities for Poverty-Related Ill Health' (2002) 16:2 *Ethics and International Affairs* 71, at 73; C. Jones, *supra* chapter 1, note 10.

[97] T.W. Pogge, *World Poverty and Human Rights*, 'Symposium on World Poverty and Human Rights' (2005) 19:1 *Ethics and International Affairs* 1; T.W. Pogge, 'Eradicating Systematic Poverty: Brief for a Global Resources Dividend' (2001) 2 *Journal of Human Development* 59; C. Jones, *supra* chapter 1, note 10.

[98] L. Doyal and I. Gough, *supra* chapter 1, note 42.

[99] Ibid., at 104.

[100] A. Buchanan, *supra* Introduction, note 30, chapter 2, at pp. 73–85.

the global structure and ensure that everyone's needs are given consideration.[101]

Duties are part of an institutional strategy to ensure that people around the world receive their basic health entitlements and profit from global equality of opportunities. Thus, a clear distinction must be made between duties of justice and acts of benevolence or charity.[102] O'Neill sets forth an interesting perspective using the Kantian universal maxim[103] to defend far-reaching justice obligations to individuals and to state that actions, policies, and institutions not be based on fundamental principles of coercion or deception within states and across borders. O'Neill asserts that there is an obligation to help, but denies that this obligation corresponds to a human right to be helped.[104] This approach fails to generate any positive duties to help those whose fundamental interests need protection through positive action. According to this approach, only non-deception and non-coercion would qualify as required conditions for justice, since positive obligations to assist others would seem to fall into the realm of beneficence rather than justice.[105]

Moreover O'Neill's approach presents only a limited perspective on obligations. In fact, as argued above, the ideal of global distributive justice involves direct help to the individuals in need of health care, resources, and technologies to bring them to a level where they can enjoy equality of opportunities. Duties

---

[101] This discussion on duties and obligations calls into question the very system of exclusive property rights which allows extensive control over knowledge, goods and technology that, if otherwise available, would be crucial in meeting individuals' basic needs and rights. With respect to property rights, individuals generally have the right to dispose of their own property as they wish. It seems that the theoretical foundation of the legal property rights needs to be questioned in our process to establish a duty to global distributive justice. This will be addressed later in Chapter 3, when we analyse the existing intellectual property system and its compatibility with distributive justice ideals. For more on control granted by property rights, see H.T. Engelhardt Jr. *The Foundations of Bioethics* (New York: Oxford University Press, 1986) at pp. 342–343.

[102] In this perspective, Loriaux draws a clear distinction between a 'duty of justice' and a 'duty of beneficence', and argues that distributive justice requires a higher degree of enforceability and responsibility than beneficence and that therefore, the current world order gives us strong reasons to recognise a global duty of justice to help the poor satisfy their basic needs. For more on this point see S. Loriaux, 'Beneficence and distributive justice in a globalising world' (2006) 20:3 *Global Society* 251.

[103] Kant's categorical imperative asks to 'act only according to that maxim by which you can at the same time will that it should become a universal law': I. Kant, *Foundations of the Metaphysics of Morals,* 2nd edn, tr. Lewis White Beck (London: Collier Macmillan, 1990) at p. 39.

[104] O. O'Neill, *supra* chapter 1, note 16, at 70; O. O'Neill, *supra* chapter 1, note 44, at 164.

[105] O. O'Neill, *supra* chapter 1, note 58, at 119–120.

of benevolence or charity are often proposed as forms of positive obligations but are not sufficient. In fact, the obligations of justice proposed here are more demanding: they require more sustained efforts, greater sacrifices, enforceable commitments from the identified duty-bearers, and the reorganisation of global institutions. Such obligations were proposed more than 30 years ago in the UN General Assembly's *Declaration on the Establishment of a New International Economic Order*,[106] which required developed countries to work toward eliminating the ever-widening gap between rich and poor countries, and to restructure the world's economic system in order to promote the economic advancement and social progress of all people. As we know, those obligations have not been fulfilled, but some actors have instead attempted to propose random and sporadic philanthropic initiatives to replace them. For example, we have seen several multinational companies offering free medication to specific countries: sometimes as an act of charity in cases of extreme health crisis, sometimes more as a result of negotiations to avoid the production of generics with the system of compulsory licenses, and other similar initiatives. Besides their sometimes doubtful safety and long-term effectiveness[107] such initiatives are not a sustainable solution, and their theoretical foundations are open to criticism. In fact, the initiatives described above can contribute to the flawed characterisation of the nature of the responsibilities of the more privileged, disguise their abusive behaviours, and obstruct in-depth institutional reforms.[108] Since they are freely and voluntarily performed, no one can claim a right to such acts as a consequence of justice. Also, no one can be forced to perform such charitable actions as generosity is voluntary and property rights over resources give owners total freedom over their management and disposition. Acts of charity that result from some sense of kindness and compassion are thus not the appropriate vehicle for achieving justice. They can be defined as moral obligations, but should not be invoked to replace legal positive obligations and duties. In fact, beneficence allows agents to withdraw from discretionary charity endeavours at any time; they can also help very limited numbers of individuals globally, which would leave many people in need of health care and resources. From a global distributive justice

---

106 GA Res. 3201 (S-VI), 2229th plenary meeting, 1974.

107 P. Berckmans et al., 'Inappropriate Drug Donations in Bosnia and Hezergovina, 1992–1996' (1997)1842:5 *New England Journal of Medicine* 337; WHO, *Essential Drugs and Medicine Policy, Guidelines for Drug Donation,* Interagency Guidelines, revised in 1999, WHO/EDM/PAR/99.4, Geneva, 1999.

108 Oxfam, *Patent Injustice: How the World Trade Rules Threaten the Health of Poor People* (Oxfam briefing paper London, 2001, online on the website of Oxfam. http://www.oxfam.org.uk/what_we_do/issues/health/downloads/patentinjustice.pdf (accessed 26 January 2009); C. Barry and K. Raworth, 'Access to Medicines and the Rhetoric of Responsibility' (2002) 16:2 *Ethics and International Affairs* 57.

perspective, this would be unacceptable, whereas from a beneficence perspective, such inequalities between individuals are simply unfortunate.[109] Indeed, Thomas Nagel makes a similar point when he indicates that 'aid should not be regarded as a voluntary contribution of a portion of a state's own wealth, but rather as a transfer of wealth required to redress distributive injustice'.[110]

2.2.3.3 Who should act?

Another important aspect of obligations and duties is the identification of duty-bearers and agents of justice. After the recognition of clear obligations, the next step is to identify who should be required to act. The purpose of this section is not to proceed in the identification of specific obligation-holders, but rather to propose generally what kinds of actors or groups should be handling such duties. Individuals, states, and other institutions and organisations may thus all be responsible for performing given duties. However, it is safe to say that, in general, collective coordination of duties will often provide better and more effective rights protection and aid provision than individuals acting alone in a disorganised manner. Tangible and definite obligations have to be supported by institutions and cultures that embody coherent and effective allocations of obligations.[111] This is not to say that citizens of powerful developed countries cannot be found directly accountable for the production of poverty and health deficits for which the governments they elected are often responsible as a result of their democratic self-interested choices in the fields of international politics and trade. As 'participants' to injustice, we all have a responsibility to people affected by such injustice. This leads us to say that inequalities in health are unjust if they are the result of unjust (international) social arrangements.[112] Therefore, individuals, as participants in the existing global institutional system, could have the responsibility of not cooperating in the imposition of unjust institutional schemes and of instead promoting reforms and establishing just institutional arrangements. The institutions thereby created would be accountable for granting efficient protection to individuals'

---

109 L.M. Fleck, *supra* chapter 1, note 6.

110 T. Nagel, 'Poverty and Food: Why Charity is Not Enough' in P.G. Brown and H. Shue (eds), *Food Policy* (New York: The Free Press, 1977) 54, at p. 57.

111 O. O'Neill, 'Public Health or Clinical Ethics: Thinking beyond Borders' (2002) 16:2 *Ethics* and *International Affairs* 35, at 39. See also: A.E. Eckert, 'Obligations beyond national borders: international institutions and distributive justice' (2008), Volume 4, No.1, *Journal of Global Ethics*, at p. 67.

112 This affirmation is rooted in Rawlsian justice and based on the idea that, in justice as fairness, the society is formed by a *fair system of cooperation* between free and equal persons (at the national level). For a more detailed discussion of this notion: J. Rawls, *supra* Introduction, note 49, at p. 311 et seq; F. Peter and T. Evans, *supra* chapter 1, note 6, at p. 28.

interests and human rights. As discussed in the first part of this chapter, as things stand now, states may well be the main agents for fulfilling duties regarding equality between individuals in the field of health and genetics. In fact, as we have seen, in the face of the actual institutional inability of the international structure to perform distributive justice duties, states are major representatives of political reality; as such, they can foster special bonds of sentiment, identity, and obligation. Such characteristics can be taken into account if the individualistic theoretical basis of justice is acknowledged and respected and the needs and interests of the others affected by those distributive mechanisms are also considered.[113] As clearly put by Pogge, I believe that: '... radical inequality can be avoided and economic human rights securely maintained within a global system of states'.[114]

Many different actors are involved in the production and distribution of genetic knowledge, technology, and resources. Of course, not all states are in a position to fulfill obligations and duties generated by individuals' interests in better access to health for equality of opportunities. Charles Jones responds to this problem by suggesting that states with more than enough resources, wealth, and technology could have a duty to redistribute such goods to more deprived states to help them meet basic and important needs with distributive justice.[115] This could be accomplished through the establishment of institutional arrangements integrating other non-state agents involved in genetics, like multinational companies, universities, and everyone who elects governments who are unlikely to self-regulate toward this end. Shue identifies those who have a primary duty to aid (*the affluent*) as the ones who spend a lot in the satisfaction of preferences as compared to rights fulfilment.[116] Nonetheless, exploring the details of the potential political actions for the assignment of duties to the affluent in this context would be outside the scope of this chapter.

The next two subsections introduce two types of obligation that have been suggested as normative grounds for a global distribution of benefits and resources.

2.2.3.4 Rawls' duty of assistance

As discussed above, Rawls' theory of justice as fairness establishes principles of distributive and egalitarian justice that apply only within individual states,

---

[113] T. Christiano, 'Democracy and Distributive Justice' (1995) 37 *Arizona Law Review* 65–72.

[114] T.W. Pogge, 'Severe Poverty as a Violation of Negative Duties', *supra* note 91, at 59.

[115] C. Jones, *supra* chapter 1, note 10, at p. 70.

[116] H. Shue, *Basc Rights*, *supra* note 63, at p. 119.

which represents a circumscribed context of social cooperation from which rights and duties arise. Even though I have already provided a critique of this aspect of Rawls' theory of justice earlier, it appears important to return to it here as one of the main propositions for distributive actions. Rawls' ideal of justice requires fair equality of professional opportunities between individuals with similar skills and talent and accepts inequalities as long as they are to the greatest advantage of the least well-off members of society. However, Rawls does not subscribe to the same standards of justice in the global context. Instead, he argues for a duty of assistance toward burdened societies, as stated in principle 8 of the *Law of Peoples*: 'peoples have a duty to assist other peoples living under unfavorable conditions that prevent their having a just or decent political and social regime.'[117] This system of transfers and mutual aid aims at bringing those societies above a minimal threshold where it becomes possible for them to satisfy their people's needs with just and decent domestic institutions. It does not, however, impose any restriction on the distribution schemes that should govern the global structure of internally well-ordered societies. According to this view, global inequalities in distribution, poverty, and wealth are meant to be dealt with internally by each domestic structure, which leaves the global level unaddressed. Those who oppose endorsing a global distributive mechanism toward the least fortunate indicate that it would ask too much on the part of those countries that are more organised, careful, and productive and that behave and invest more responsibly and reasonably.[118] This is why they call for a well-circumscribed duty of assistance, with a clear objective and cut-off point instead of the establishment of international institutional mechanisms for distribution. The duty of assistance seems to derive from the importance of expanding the Society of Peoples to include every society in the world; in so doing, Rawls completely avoids the notion of global distributive justice. Such duty of assistance does not entail the obligation to reduce inequalities among individuals living in societies with different endowments of natural or human resources, different histories, or different cultures.[119]

Rawls' duty of assistance is not sufficient and can be criticised on various grounds. This duty to aid is not qualified as a collective responsibility of well-off societies and there is no mention of a right of the less-advantaged to receive any benefits. Rawls does not provide much detail on the scope of this potential duty, making it seem more like a vague 'duty of charity' than an obligation of justice as described earlier.[120] By not endorsing global distributive principles

---

117 J. Rawls, *supra* chapter 1, note 11, at pp. 107, 115–119.
118 A. Buchanan, *supra* Introduction, note 30.
119 C.R. Beitz, *supra* Introduction, note 49.
120 A. Buchanan, *supra* Introduction, note 30.

for the reason that it would result in some states bearing some costs that arise from decisions made by others,[121] it emphasises the negative aspects of the internal structure, culture, and tradition of less-advantaged countries. As discussed above, and bears repeating, the ability of a society to reach a higher stage of development and address internal inequities is not only a result of voluntary economic and political choices at the domestic level, but is also a direct consequence of its natural resource endowments and situation in the international political economy.[122] Indeed, Rawls' duty of assistance does not protect poor societies from the international terms of cooperation imposed through negotiations that are greatly affected by the unbalanced bargaining power that marks the basic global structure. His account is misleading since it emphasises lack of assistance to the deprived rather than questioning the justness of the global order that is imposed by the most wealthy and powerful.[123] Rawls seems to recognise this issue as he mentions that the 'unjustified distributive effects' of cooperative organisations need to be corrected, but he does not go further, instead endorsing a duty of assistance that does not allow such correction.[124] Rawls' fear over 'open-ended' redistributive initiatives should not prevent any sort of distributive commitment, since limited egalitarian principles can be adopted to constrain inequalities in specific spheres of importance, such as health and basic needs.[125]

Furthermore, the cosmopolitan focus of this book demands that we consider each individual as a standard of moral concern, but Rawls' duty of assistance does not permit such consideration. It instead allows the major health and welfare gap between individuals living in different countries of the world to persist. By emphasising the responsibility of burdened societies as a rationale for a limited and simple duty of assistance, Rawls does not acknowledge that most individuals living in those societies have no power whatsoever over poor investments. In fact, the conduct of an irresponsible country is often dictated by a few elites who act alone, without their population's assent or participation. It is therefore difficult to conceive how previous, current, and future generations could be held responsible for those choices.[126] As we clearly cannot rely on Rawls' duty of assistance to support our moral framework of global distributive justice, some have instead proposed to extend Rawls' difference principle to the global context as a basis for obligations of justice.

---

121 J. Rawls, *supra* chapter 1, note 11, at 74–77 and 105.

122 A. Kuper, *supra* chapter 1, note 96.

123 T.W. Pogge, 'Cosmopolitanism and Sovereignty', *supra* chapter 1, note 18.

124 J. Rawls, *supra* chapter 1, note 11, at 43 and 115.

125 For example, refer to A. Buchanan, *supra Introduction,* note 30.

126 C.R. Beitz, 'Social and Cosmopolitanism liberalism', *supra* Introduction, note 47.

2.2.3.5 Rawls' difference principle applied globally

In his *Theory of Justice*, Rawls argues that rational individuals who would position themselves behind a 'veil of ignorance',[127] after ensuring equal distribution of basic liberties, would choose two basic principles to govern distribution in their institutions: equality of opportunity (professional) and the difference principle (difference in social primary goods such as wealth, power, income, and the social base of self-respect are to be justified only if they make everyone better off and are to the greatest advantage of the least well-off). This test aims at identifying principles that will promote the good of individuals as equal moral entities. Rawls uses this contractual test to develop our traditional notions of moral obligation, express the inherent moral standing of persons, and negate differences in bargaining power.[128]

The difference principle gives rise to a duty to eliminate existing inequalities in order to comply with an egalitarian principle of justice as opposed to satisfying claims to reach certain limits, beyond which limits no more equalisation would be required. The difference principle calls for transfers on purely egalitarian grounds. Endeavours of high importance, the realisation of special values and the meeting of basic needs are not taken into consideration by the application of the difference principle. Its aim is mainly to minimise unjustified inequalities.

Rawls developed the difference principle in the context of his theory of domestic justice as fairness and, in his opinion, it should only apply to the distribution of wealth and income within societies. However, as demonstrated earlier, in the actual global structure, boundaries do not establish and limit the scope of social cooperation; they should therefore not restrict associated social obligations. I believe that the statist perspective of the world has lost its normative significance due to the rise of global economic interdependence. Consequently, distributive responsibility of states should simply represent a continuation of our general duty of justice at the global level. According to this view, the difference principle adopted in the domestic context would also be selected as a standard of justice in the global context where, due to an extended veil of ignorance, issues of citizenship would not be taken into account.[129] As a result, a just global distribution of social primary goods that could be distributed by social institutions (such as income, wealth, powers) would have to maximise the absolute position of the least privileged individuals and societies of the global order.

---

[127] A fictional position where they would know neither what type of society they would live in nor their place and position in this society.

[128] W. Kimlicka, *supra* chapter 1, note 39.

[129] C. Jones, *supra* chapter 1, note 10.

As discussed above, in Rawls' view, the less advantaged position is determined in terms of individuals' possession of primary social goods, as opposed to natural primary goods like health, opportunities, and natural talents. Therefore, with Rawls' theory, health is not a criterion for determining the position of an individual, but wealth and income are. Some have suggested that inequalities in health that are influenced by wealth should also be justified by the difference principle; others endorse the premise that natural primary goods and endowments should also be taken into consideration in the definition of the least-privileged.[130] If we take those views and consider that health inequalities should not just be analysed with the principle of equality of opportunity but also pass the difference principle test, it would mean the prioritisation of the least-favoured groups in society in terms of health conditions or the improvement of the health of the poorest members in society to agree to health inequalities. The application of this reasoning might lead to a desirable outcome in our process of aiming at a global distribution of the genetic-innovation benefits. If we take the criterion for the identification of the least privileged globally to be poor health due to class inequality, this group would likely comprise a majority of individuals from developing countries. It would thus mean that global inequalities in the distribution of the genetics-research benefits and the investment of research funds could be justified under the difference principle if they would improve, even very slightly, the situation of those living in developing countries. Is this really what we are aiming at? Would this kind of solution allow every individual's genetic needs to be met globally?

The scope of the underlying elements of the difference principle remains very unclear. It gives no precise indication of the conditions to meet in order to be part of the least-privileged group and does not detail what is required to attain the *greatest benefit for the least-advantaged group* threshold. Therefore, in the case under study, it appears that the application of the difference principle could leave many sick individuals (for whom access to genetic innovations could make a difference) outside of the redistribution scheme depending on the interpretation of the difference principle. In fact, the reason transfers to the least privileged are required is not to enable them to realise specific values or to meet specific crucial needs, but to minimise unjust inequalities.[131] For example, individuals affected by health problems in a difficult socio-economic context, but not necessarily falling within the world's least privileged group (for example, people living in middle income countries), could be left out of

130 S. Marchand, Daniel Wilker and B. Landesman, *supra* note 44.
131 W. Hinsch, *supra* chapter 1, note 33.

the distribution process.[132] In fact, a *maximin* principle like the difference principle could justify giving priority only to the least privileged as opposed to a focus on all class inequalities in health.[133] Also, the 'worse off group' could be interpreted as those who have the most urgent and serious health conditions without any reference to socio-economic conditions. Such an interpretation could also leave many individuals in need of the benefits of genetic therapies outside of the realm of distribution.

The application of the difference principle could also allow major differences in access to genetic services and technology to persist. In fact, the standard of the difference principle does not require equality, but only an absolute improvement of the situation of the least well-off to justify inequalities. It does not ensure that the basic needs of the least well-off will be met and that they will be able to flourish as human beings. Therefore, major global inequalities in access to genetics could be tolerated given the observation of an absolute improvement of the situation of the globally less-privileged group. In fact, the application of the difference principle can result in the reward of productivity and wealth and the endorsement of a hierarchy between individuals and societies, provided only that the less-privileged agents also receive some benefits. For example, limited access to available genetic technologies by some very poor populations in need could be seen as sufficient improvement in the actual situation of the global least-privileged, which could consequently justify unequal distribution and access to genetic benefits more generally by the affluent. As Shue demonstrates, this aspect of Rawls' theory does not provide everyone the means to keep their heads above water; what it does is merely allow people to 'continue to drown but with less and less water over their heads'.[134] It is thus safe to say that the economic order created by Rawls' theory is characterised by free bargaining and the improvement of the wealthier societies' position. As such, it is open to criticism for the above mentioned reasons, even if it prevents the most disadvantaged from falling below a certain minimum threshold.

Since a number of global inequalities in health could be left unaddressed by the application of the distributive duties generated by the difference principle, it is not part of my ideal approach as it stands now. This takes me back to our concept of justice in health and to the right to equality of opportunities. As argued above, the right to equality of opportunities and the corresponding duties of the affluent to give aid provide a better rationale for distribution as

---

132 T. Nagel, *Mortal Questions* (Cambridge: Cambridge University Press, 1979) at pp. 122–130; A.D. Williams, 'The Revisionist Difference Principle' (June 1995) 25:2 *Canadian Journal of Philosophy* 257, at 280.

133 S. Marchand, D. Wilker and B. Landesman, *supra* note 44, at p. 461.

134 H. Shue, *Basic Rights supra* note 63, at p. 128.

compared to a duty of assistance, charity, or beneficence. Moreover, Rawls gives priority to this notion before addressing acceptable inequalities with the difference principle. This means that ensuring individuals are brought to an appropriate level of normal functioning, to make sure they can profit from available opportunities, should not be compromised by inequalities permitted by the difference principle. In other words, attaining normal functioning to be in a position to benefit from equality of opportunities is a priority, and is therefore supposed to be protected against interferences from inequalities persisting with the application of the difference principle.

However, as mentioned earlier, since a just system involves many requirements in various areas and because of the reality of limited health resources for unlimited health needs, real and universal equality of opportunity remains out of reach (for now) at the non-ideal level. Although I focus on an ideal framework of justice, I can say that intermediate standards will be needed at first, to allocate scarce resources to our health justice ideals. As health can be influenced by social primary goods like wealth and income, an intermediate distributive standard could take the form of a modified difference principle with a special focus on basic needs, as proposed by Doyal: 'Rawls's difference principle should be expanded to state that inequalities will only be tolerated to the extent that they benefit the least well-off through leading to the provision of those goods and services necessary for the optimisation of basic need-satisfaction'.[135] Similarly, others have proposed, instead of focusing on the least well-off, that health-resource distribution should favour individuals who are below a threshold level of health. This system would not require achieving equality in health, something that appears impossible given the current high level of health in industrialised countries, but would require that no one remain below this acceptable minimum level of health. This global threshold would be defined in light of existing medical technology.[136] These propositions avoid the troubling possibility of having to trace a line between the absolute worst-off and others who are not part of the least-healthy category but should nevertheless have their basic health needs taken into account when undertaking distribution of health and genetic resources. In referring to a basic need for health and a threshold level of health, these new versions of the difference principle would secure, in priority, a distribution of health and genetic resources for the people who need them to meet their basic needs.

As my basic claim is to argue for more equitable distribution of resources to vulnerable populations globally, taking need and normal functioning into account, these propositions are a good starting point to meet my goal. Ideally,

135 L. Doyal and I. Gough, *supra* chapter 1, note 42, at 132.

136 A.K. Acharya, *supra* Introduction, note 6; T. Nagel, *supra* note 110, at p. 125.

however, I believe that just distribution should go one step further and consider needs and normal functioning in their broader context, as influencing the fair range of opportunities that should be available to every individual.

CONCLUSION

In this chapter, I have provided an analysis of the specificity of health as the central part of my framework of distributive justice. I first emphasised the importance of health and genetics in normal functioning and then the role of normal functioning in allowing individuals to profit from equality of opportunity. After having established that fair distribution was required in this field, I addressed if and how it should be undertaken, analysing correlative rights and obligations. As a result, I was able to build a solid argument for the use of distributive justice mechanisms to solve avoidable health inequalities and foster equitable access to the benefits arising from genetics.

More generally, I dedicated the first part of this discussion to the construction of a global distributive justice framework to serve as a basis for more equitable access to health and genetic care, benefits, and resources. In this ideal moral scheme, every individual's health interests receive equal consideration in the pursuit of equal opportunities.

This ideal conception of justice could be criticised on the ground that it demands unrealistic health standards that cannot be incorporated into the current global order. This critique emerges from an institutional conception of justice that starts from the actual world structure, characterised by existing states, territorial boundaries, and strong power differences. My conception does not try to change or eliminate it, but instead aims to discover how we can continue to support it with a different focus, one that takes principles of justice into account.[137] To justify their focus on the national structure, some argue that the institutional arrangements that can help develop the basis for cooperation, that can allow political coercion and initiate shifts in the distribution of wealth and power, are often provisional and insufficient at the international level. They are even talking about democratic deficits and lack of accountability on the global scene, due, in part, to state differences and mostly to the growing

---

137 Such a view has been qualified as an institutional form of an ideal theory since it does not take non-ideal conditions for granted, but instead aims to find justifications for existing institutions, taking ideal circumstances into account. It is different from the ideal theory I argue for. In fact, my primary goal is not to justify the existing institutions, but to determine how best to meet our global distributive justice ideals in health, preferably but not necessarily within the actual institutional structure. M. Blake, *supra* chapter 1, note 53, at 262–264.

presence of powerful non-state actors. The existing state system and international order, their capacity to integrate structural changes and interact together, and the main actors in charge of shaping and controlling them are important aspects of the actual political reality that need to be considered. Another critique of global distributive justice could come from the fact that enforcement mechanisms for compliance with minimal international redistributive policies are lacking. For example, the United Nations has failed to convince the affluent to supply as little as 0.75% of their gross national products for international development initiatives.[138] Therefore, the practical application of the concept of ideal justice will inevitably be constrained by the institutional reality of the world.[139] Indeed, those who express scepticism about an ideal account of justice believe that it might be incompatible with reality, more specifically with the political account of how the present system of international economic inequality came into existence, whose interests it serves, and how it can effectively be adapted or changed to serve the interests of the poor and deprived.[140] As Onora O'Neill states, 'knowing that some distribution (equal, *maximin*, or whatever) of resources, or of health care, would be *ideally* just does not take us far toward knowing who should do what for whom in order to work toward that distribution'.[141]

I believe in the relevance of an ideal theory of justice. As mentioned at the very beginning of this first part, I consider that it is of the utmost importance to get a sense of first, what we are setting aside when agreeing to non-ideal conditions and second, how we can envision social reforms. To this effect, Thomas Pogge states:

---

138 *Towards Accelerated Development – Proposals for the Second United Nations Development Decade* (United Nations Publication, E.70.II.A.2, 1970); UN Committee for Development Policy, The role of the Committee for Development Planning (CDP) in the formulation of the United Nations International Development Strategies (for the 1970s, 1980s and 1990s) and a Summary of CDP's Main Recommendations for those Strategies, CDP/2000/PLEN/10, 20 March, 2000, online on the UN website: http://www.un.org/esa/policy/devplan/cdp00p10.pdf (accessed 27 January 2009).

139 For example, the structure and functioning of TRIPS is a reality of the global order and could have negative impact on developing countries' welfare and development. We will be coming back to this specific system later in the book, but for a discussion on this point refer to J.H. Reichman, *Implications of the Draft TRIPS Agreement for Developing Countries as Competitors in an Integrated World Market*, UNCTAD Discussion Papers no. 73 (UNCTAD/OSG/DP/73, November 1993); A.S. Oddi, 'TRIPS: Natural Rights and a Polite Form of Economic Imperialism' (May 1996) *Vanderbilt Journal of Transnational Law* 29.

140 C. Brown, *supra* chapter 1, note 1, at 180.

141 O. O'Neill, *supra* note 111.

> Realism hardly requires that the principles of justice conform themselves to the prevailing sordid realities. We don't feel justified to give up our ideals of domestic justice or personal honesty just because we despair of achieving them fully. We cannot reasonably demand of moral principles that they vindicate the status quo. All we may ask is that a conception of justice provides a criterion for assessing our global order that allows us to choose from among the feasible ... avenues of institutional change and thus specifies our moral task – gradually to improve the justice of this order.[142]

By focusing on establishing ideal principles of justice in health, we acknowledge that the existing order is not completely static and unalterable. Indeed, the global order paradigm is less than 60 years old, and this might mean that 'changes can be achieved through human agency in response to changing times'.[143] This is especially true in the sphere of health, where it has become quite clear that individual and isolated actions and initiatives for improving human health are not providing the sustainable changes required at the global level. As such, one of the biggest challenges is a shift in perspective from economic self-interest to growing solidarity and a shared spirit of mutual caring.[144] There is no reason to think that such changes of perspective would be impossible. Indeed, it is interesting to observe how much institutional change has been achieved over a relatively short period of time. Many of the most influential institutions in the world have appeared in the last 50 years. In fact, most trans-national corporations, international organisations, banks, and development agencies are new types of actors in the ever-changing global picture. However, the type of transformation we are aiming for is unlikely to take place automatically and voluntarily in a world mainly driven by market powers and self-interest.[145] This is one of the reasons why I do not argue for

---

142 T. Pogge, *Realizing Rawls* (Ithaca: Cornell University Press, 1989), at p. 260.

143 M. MecGwire, 'The Paradigm that Lost its Way' (2001) 77:4 *International Affairs* 777, at 793.

144 R. Rorty, *Contingency, Irony and Solidarity* (Cambridge: Cambridge University Press, 1989); S.R. Benatar, A.S. Daar and P. Singer, *supra* note 1, at 122.

145 Many have proposed practical initiatives to reunite the ideal theory with the non-ideal reality. I will come back to some of them in the conclusion, to set the basis for further discussion. For example, some have proposed a very egalitarian solution to the issue of access to health care resources and genetic services and technologies, arguing that their use should be banned unless they can be made available to everyone who needs them. This argument is based on the fundamental equality of persons that should entail equal opportunity, respect and treatment. Such position has been endorsed by the Council on Ethical and Judicial Opinions of the American Medical Association with regard to the availability of genetic enhancement of foetuses and embryos. A. Gutmann, *For and Against Equal Access to Health Care, in Securing Access to Health Care, ed. President's Commission for the Study of Ethical Problems in Medicine and Biomedical and Behavioural Research* (Washington, DC: US Government Printing

starting from scratch and ignoring the existing institutional order, agreeing instead to work with some of the institutions already in place when major changes can be undertaken to reflect the health equity concerns that constitute our argument.

The purpose of the second part of this book is to present two important existing international normative systems and to determine whether their structure and functioning adequately account for, and balance, the many values of our global distributive justice framework in facilitating a future redistribution of potential benefits in the field of genetics. If need be, distributive justice will guide the critique of the actual framework and the reconstitution of the ground rules that should regulate property, cooperation and exchange, as well as conditions of production and distribution.[146]

Office, 1983) at 53; American Medical Association, Council on Ethical and Judicial Affairs, 'Ethical Issues Related to Prenatal Genetic Testing' (1994) 3 *Archives of Family Medicine* 633, at 640–641. Others suggest a multilayered institutional scheme in which government authority, instead of being concentrated at the state institutional level, would be dispersed among different political units, and not necessarily constrained by existing historical borders. Therefore, spreading authority over different units could reduce the incidence of poverty and oppression, factors that too often shape the actual state of the world order. In a sense, the adoption of political agencies that properly control different spheres of human action, not all territorially-based, is an alternative to both a world state and a state-dominated system. T.W. Pogge, 'Cosmopolitanism and Sovereignty', *supra* chapter 1, note 18. Thomas Pogge also proposes a transfer of one percent of the affluent classes' gross domestic product to worse-off states, as well as a global resource tax – two options that have been criticised and which in any case would be interesting to investigate further. T.W. Pogge, 'Eradicating Systematic Poverty: Brief for a Global Resources Dividend' (2001) 2 *Journal of Human Development* 59; T. Pogge, 'A Global Resources Dividend' in D.A. Crocker and T. Linden (eds), *Ethics of Consumption: The Good Life, Justice and Global Stewardship* (Lanham, Md.: Rowman and Littlefield, 1997).

146 T.W. Pogge, *supra* chapter 1, note 18, at 97.

PART II

Some normative tools for distribution in health

HOW DOES OUR NORMATIVE THEORY OF DISTRIBUTION TRANSLATE INTO POSITIVE LAW NOW? THE CASES OF INTELLECTUAL PROPERTY AND INTERNATIONAL HUMAN RIGHTS LAW

The first part of this book established clear normative landmarks to assess how equitably the world distributes health and genetic benefits. It set out an approach that gives equal consideration to everyone's basic health needs, adopting a cosmopolitan approach where individuals are treated with the same consideration regardless of their citizenship or geographical location. I then reflected on the universal importance of health, its vital role in normal functioning and in the pursuit of equal opportunity for all. This analysis led us to establish a global distributive justice framework in health which supports my argument for equitable access to the benefits arising from genetic advances, taking basic health needs and opportunities into account.

Now that I have built those theoretical foundations, I can begin discussing how this normative approach to distribution translates (or not) into positive law. To this end, I will identify and analyse the main obstacles to legal compliance with global distributive justice in health and genetic research. This will be the main goal of the second part of this book.

Although many spheres of law can be applicable and useful when dealing with the different issues emerging from the development of genetic science, I decided to concentrate on the two that appeared most relevant and important to addressing distributive justice issues in health: intellectual property law and human rights law. Not only are these two legal systems discussed extensively in the literature, they are also very important in the global normative picture in their own different ways – especially in terms of access and equity issues.

Therefore, the next two chapters will aim to present and analyse the international intellectual property and human rights systems, assessing their underlying philosophy, construction, and application with the precise benchmarks of justice established in our theoretical framework. This analysis will help us identify important weaknesses of these legal frameworks and to realise that, although of very different nature, they are both greatly influenced by powerful agents and market factors which undermine their focus on justice and equity issues.

3. International intellectual property law: a first tool?

INTRODUCTION

The aim of this chapter is to determine whether the foundation, structure, and purpose of the existing intellectual property (IP) law system, and especially patent law, assists or hinders the realisation of global distributive justice in health and genetics. I will commence with a brief introduction to intellectual property rights, in particular, patent law and its application to genetics. The second part of this chapter aims to present and provide a succinct analysis of some of the main theoretical foundations brought forward for justifying property rights on intellectual inventions. The third section assesses the patent system in referring to considerations of distribution, equality, and justice. An evaluation of the patent system by reference to the standard of access (global access to resources, availability and affordability of products and services) will be provided in order to establish whether the international patent system can serve the purpose of global distributive justice.

INTRODUCTION TO INTELLECTUAL PROPERTY

Property rights are used as legal and political tools to help ensure social order, structure, and harmony in communities. They translate the connection between property holders and non-holders into enforceable legal rights. Through this system, objects of property can be viewed as articles that can be traded in the market, providing property owners with some degree of economic power.[1] In this sense, *property institutions fundamentally shape a*

[1] C. May, 'Unacceptable Costs: The Consequences of Making Knowledge Property in a Global Society' (2002) 16:2 *Global Society* 123; A. McEvoy, 'Market and Ethics in United States Property Law' in H.M. Jacobs, ed., *Who Owns America? Social Conflict Over Property Rights* (Madison: University of Wisconsin Press, 1998) 94, at p. 99.

society.[2] Property rights can be associated both with tangible and intangible and intellectual objects. These are referred to as intellectual property rights (IPRs), and can be defined as rights in original ideas included in tangible products of cognitive effort, which give IP holders a legal right to exclude people from making use of their property in exchange for a public disclosure of the object of their right.[3]

Intellectual property has increasingly become a prevalent form of ownership and signifies a very valuable asset for many IP holders worldwide. In fact, the economic significance of IP in the global market represents hundreds of billions of dollars and is constantly growing.[4]

Although intellectual property refers to different forms of legal protection (patents, copyrights, trade secrets, trademarks and so on), some general features are common to all forms of IP. For example, the object of intellectual property is intangible, and is therefore non-exclusive; that is, it does not disappear after it has been used or shared. In other words, the possession or use of any intellectual object by one person does not prevent others using or possessing it concurrently. Hence, in order to enhance the dissemination of ideas (copyrights and patents) or to encourage the creation of proprietary information (trade secrets), IP artificially creates scarcity. In fact, it allows holders to exclude people from using their intellectual objects even if simultaneous uses by a multitude of individuals would be possible without additional cost or risk of overexploitation.[5] Another particularity of intellectual property that differentiates it from material property is the temporal limits associated with the rights awarded. In most cases, IPRs are granted by states for a fixed period, after which the objects of IP become freely available to the community as part of the public domain.

The protection awarded to intellectual property in the field of human genetics is mainly established in patent and copyright systems.[6] Before addressing

---

2 E.C. Hettinger, 'Justifying Intellectual Property', in A.D. Moore (ed), *Intellectual Property: Moral, Legal, and International Dilemmas* (Maryland: Rowman & Littlefield, Lanham, 1997) 17, at p. 27; J. Boyle, *Shamans, Software and Spleens: Law and the Construction of the Information Society* (Cambridge and London: Harvard University Press, 1996).

3 J. Hugues, 'The Philosophy of Intellectual Property' (1988) 77:13 *Geo. L. J.* 287, at 294–296.

4 J. Boyle, *supra* note 2, at p. 121; L.G. Thurow, 'Needed: A New System of Intellectual Property Rights' (Sept–Oct 1997) *Harv. Bus. Rev.* 95, at 96–97.

5 E.C. Hettinger, 'Justifying Intellectual Property' in P. Drahos (ed), *Intellectual Property* (Aldershot: Dartmouth Publishing, 1999) p. 117.

6 Trade secrets can also be used to protect the confidentiality of important and valuable business information and encourage the production of proprietary information by offering protection against its misappropriation. The same subject matter protected

patent issues, which will be the focus of this chapter, it is necessary to briefly discuss the application of copyright to genetics.

There is considerable copyrightable material involved in the field of genetics. Indeed, since the outcomes of DNA sequencing are often used as an information storage base for future breakthroughs, and require substantial and lengthy further analysis, it is safe to say that there is a great informational potential and value in some unique collections of DNA sequences. Copyright law can be used to protect the value of some original compilations of results arising from genetic research including and not limited to a gene sequences database, a list of single nucleotide polymorphisms (SNPs), a diagram of the order of the fragments on some molecules and so on.[7] Since most genetic research data is collected in databases of different forms, many genetic and genomic compilations qualify for copyright protection and can therefore be subject to access and subscription fees.[8] However, as numerous other aspects

---

by trade secrets could instead be patented or copyrighted; the choice of one protection regime over another is a matter of business strategy. Trade secrets do not have to be publicly disclosed (they must remain secret), and they can last as long as they stay confidential. One problem with trade secrets, however, is that once the object of the secret is revealed, disclosed, or figured out, anyone can use, reproduce and sell the invention or process without any restrictions from the trade-secret holder. For example, if a computer hacker deciphers how to access the secret and then steals it; or if there is a leak from inside and the secret is revealed; or if someone invents, discovers, or creates the equivalent of the secret's subject matter independently, the system of trade secrets could be invoked to claim reparation from the thief for breach of security, or the employee for breach of trust and confidentiality. Nothing, however, could prevent the public from freely making use of the revealed secret. This is probably one of the reasons why the use of trade secrets is not widespread in the field of biotechnology and genetics, where keeping secrets can prove quite difficult. Since a great deal of genetic research involves the use of similar technologies, and aims to isolate and identify the same biological functions of the same genetic elements, the chances are that different research teams could, at a certain point, independently and accidentally produce identical results. Unless the object of the secret is completely novel and original in the sense that it has no equivalent in the natural world, researchers will often be capable, over time, of finding out the subject matter of a trade secret by replicating it from material existing in nature. Therefore, given that genetic research and innovation often involve huge investments in terms of time and capital, it is unlikely that the majority of inventors and authors would risk protecting their ideas through the trade secret regime. For more on trade secrets, refer to: T.A. Lipinski and J. Britz, 'Rethinking the Ownership of Information in the 21st century: Ethical Implications' (2000) 2:1 *Ethics and Information Technology* 49; D.B. Resnik, *Owning the Genome: a Moral Analysis of DNA Patenting* (Albany: State University of New York Press, 2003).

7 H. Haker, R. Hearn and K. Steigleder, *Ethics of the Human Genome Analysis* (Germany: Attempto Verlag Tübingen, 1993) at p. 112.

8 R.S. Eisenberg, 'Re-Examining the Role of Patents in Appropriating the Value of DNA Sequences', (2000) 49 *Emory L.J.* 783.

of genetic research do not give rise to copyrightable material, we must turn to another IP protection regime: patents.

In the field of genetics, patents are certainly the preferred legal system for protecting genetic inventions, investments, and benefits, due to the broad scope of application and the important financial returns they can engender. The following section will present the patent system as applied to the field of genetic development, addressing some ethical debates raised by gene patents and discussing the national and international aspects of the patent system.

3.1 THE PATENT SYSTEM

Genetic research and development is a field giving rise to substantial ethical and legal debates, among which patents are one of the most litigious issues.[9] The patent gives the inventor the right to exclude other people from using, making, importing, or selling his inventions for the duration of the patent, in exchange for wide disclosure and publication of a detailed description of his invention. In this sense, patent rights reveal a natural tension between dual roles: protection and dissemination.[10] To be patentable, an invention (products or processes) must be new (not previously patented or published), involve an inventive step (non-obvious improvement of what already exists), and be capable of industrial application (useful).

The main purpose of awarding exclusionary rights to someone for an eligible invention is to allow the patent-holder to recoup the time and funds invested in developing the invention and, subsequently, to encourage more innovation.[11] Patents raise the issue of monopoly power, as they allow patent owners to have some control over prices and productivity.[12] Patent rights thus create an exception to the principle of free enterprise, competition, and availability of information on the basis that they are meant to promote further tech-

---

[9] A. Finlay, 'Gene Patenting: Seeking Benefits for All' (2003) 82 *Reform* 52 at 53.

[10] J. Boyle, 'A Theory of Law and Information: Copyright, Spleens, Blackmail, and Insider Trading' (1992) 80 *California Law Review* 1415, at 1440.

[11] The amount of money invested in the development of an invention can be very high in certain fields of activity. These funds can be used to cover the costs of R&D, make up for ineffective trials, and get the necessary regulatory approvals. The exact amount invested varies depending on the sector and the companies, and can be as high as several hundred million dollars to develop and market a single drug. For more on R&D investments refer to: F.S. Kieff 'IP Transactions: On The Theory & Practice of Commercializing Innovation' (2005) 42:3 *Houston Law Review* 727.

[12] M. Trebilcock and R. Howse, *The Regulation of International Trade* (New York: Routledge, 1995) at p. 249.

nological progress and innovation.[13] In addition, they could prevent significant profit losses by restricting imitation.[14] In other words, some believe that the patent system, by imposing temporary restrictions on widespread use and access to knowledge, information, and ideas, has an important role to play in boosting the production of crucial knowledge and innovation in biotechnology, genetics, and health.[15] In the field of genetics, translating scientific discoveries into useful therapeutic products and services can be a long, complex, and expensive process. Additionally, the economic value of patents awarded in this field can be very high.[16] Thus, many acknowledge the potential advantages of granting patent protection in this field to stimulate investment and encourage further developments in an area with tremendous promise for global health.[17] One reason given to justify the importance of patents in genetics and biotechnology is the major difference between the costs of innovation and of imitation in these specific fields of scientific activity.[18] It is important to mention, however, that others still consider there to be very little empirical evidence sustaining this incentive theory, which furthermore refers only to a small subset of the whole of innovation.[19] We will return to this specific point throughout this chapter.

Even after more than 20 years of existence, patents on genetic material are still at the centre of a number of social policy dilemmas, particularly on the very nature of genes and their capacity to qualify as patentable material in relation to the basic patenting rules. Nevertheless, the current international legal

---

[13] E.R. Gold, 'Finding Common Cause in the Patent Debate' (2000) 18 *Nature Biotechnology* 1217.

[14] H. Grabowski, 'Patent, Innovation and Access to New Pharmaceuticals' (2002) *J. of Int'l. Eco L.* 849, at 850; M. Trebilcock and R. Howse, *supra* note 12 at 250.

[15] For proponents of this position, the benefits of patents vis-à-vis innovation and development prevail over the cost of exclusivity for society. To illustrate this point, one author refers to the example of granting patent rights over the cure for cancer, arguing that while it may sound outrageous that a company could gain exclusive rights over such a crucial matter for 20 years, without a patent inducement, it could take hundreds of years longer to find the same treatment for cancer. D.L. Burk and M.A. Lemley, 'Policy Levers in Patent Law' (2003) 89 *Vanderbilt Law Review* 1575, at 1581.

[16] For example, the value of the patent right awarded on the gene involved in producing a hormone used in kidney disease treatment (erythropoietin) is about $1.5 billion a year. L.B. Andrews, 'The Gene Patent Dilemma: Balancing Commercial Incentives with Health Needs' (2002) 2 *Hous. J. Health & Pol'y* 65.

[17] UNESCO International Committee on Bioethics, *Report of the IBC on Ethics, Intellectual Property and Genomics*, 10 January 2002, SHS-503/01/CIB-8/2 Rev.

[18] R.M. Cook-Deegan and S.J. McCormack, 'Patent Secrecy and DNA' (2001) 293 *Science* 217, at 217.

[19] S. Macdonald, 'Exploring the Hidden Costs of Patents' in P. Drahos and R. Mayne (eds), *Global Intellectual Property Rights, Knowledge, Access and Development* (New York: Palgrave Macmillan, 2002) 13.

consensus is that isolated human genetic material is patentable. In fact, claiming strong proprietary rights in genetic material has been a widespread practice since the early 1980s[20] and, in the majority of countries, none of the ethical and social concerns voiced by various stakeholders have changed this permissive approach. Instead, we have witnessed an explosion of human genetic patent applications in terms of quantity and diversity in the major industrialised countries.[21] As of September 2008, over three million genome-related patent applications had been filed worldwide.[22]

Awarding patents in human genetics and especially gene patenting has nevertheless given rise to numerous ethical debates.[23] The main ethical issues generated by patents in genetics will be analysed and re-examined at more length and detail throughout this chapter. For the moment, we can say that many ethical issues arise from the very dichotomy of patent rights that are meant to promote a balance between appropriation/protection (the inventor's rights) and dissemination (the community's rights). Some important issues arising from genetic patents relate to human dignity; access to research, products, and services;[24] and to the suitability of granting exclusive property rights

---

[20] D.B. Resnik, 'The Human Genome: Common Resource but not Common Heritage' in *Proceedings from a Frontis Workshop on Ethics In the Life Sciences*, Wageningen University, 2004, chapter 13, 197, at 203.

[21] Indeed, in the field of human genetics, it is possible to patent processes like methods for isolating, purifying, cloning, multiplying, changing, examining, and manufacturing DNA. In addition, genetics patents can also be granted on resources e.g. DNA sequences, genes and their end point markers, expressed sequence tags (ESTs) isolated and purified, spliced into recombinant vectors, or introduced into recombinant cells under laboratory conditions. Since DNA sequences are both molecules and information, it is important to determine what is actually covered by exclusive patent rights. Patent rights over genetic composition of matter should give the patent-holder temporary exclusive rights over the material substance of the molecule. However, the very nature of the patent bargain imposes, in exchange for this exclusive right, the disclosure of the invention itself and of information about the nature, functioning, and properties of the invention. The patent system thus allows the public to get access to, use, and analyse this genetic information, with the correlated obligation to respect the inventor's conditions, which can sometimes be very restrictive, onerous, and demanding. For more details, refer to D.B. Resnik, 'DNA patents and human dignity' (2001) 29:2 *The Journal of Law, Medicine, and Ethics* 152.

[22] Human Genome Project Information, online HGPI: http://www.ornl.gov/sci/techresources/Human_Genome/elsi/patents.shtml (accessed 20 January 2009) 'Genetics and patenting, what are patents, and how do they work?'

[23] T. Caulfield, E.R. Gold and M.K. Cho, 'Patenting Human Genetic Material: Refocussing the Debate' (2000) 1 *Nature Reviews Genetics* 227; B.M. Knoppers, 'Status, Sale and Patenting of Human Genetic Material: An International Survey' (1999) 22 *Nature Genetics* 23.

[24] M.A. Heller and R.S. Eisenberg, 'Can Patents Deter Innovation? The

over material embodying essential information for building our common knowledge of human genetics and genomics.[25] Therefore, granting exclusive property rights over genetic material is, for some, equivalent to allowing the commodification and gradual drain of some common asset of humanity and keeping it out of access for subsequent basic research and important screening and therapeutic purposes.[26] Consequently, endorsing private exclusionary rights in such material for a few select, wealthy corporations and countries can lead to substantial health inequalities based on economic considerations. Others defend the view that temporary appropriation of such material is essential to foster subsequent scientific innovation, and that preventing genetic patents would be equivalent to promoting unreasonable use of those resources.[27] Let us examine a few of the major ethical arguments relating to the patentability of genetic material.

3.1.1 Human Genetic Material: Patentable Substance?

There has been an ongoing debate as to whether genetic material should qualify more as discovery or invention.[28] Today, patents are generally conferred on some isolated and purified genetic material, on the basis that human intervention was required to take it from its natural stage and bring it to its new stage.

Anticommons in Biomedical Research' (1998) 280 *Science* 698; M.R. Henry et al., 'DNA Patenting and Licensing' (2003) 297 *Science* 1279.

[25] S. Sell and C. May, 'Moments in Law: Contestation and Settlement in the History of Intellectual Property' (Autumn 2001) 8:3 *Review of International Political Economy* 467, at 474.

[26] J. Boyle, *The Public Domain: Enclosing the Commons of the Mind* (New Haven: Yale University Press) 2008.

[27] German National Ethics Council, *Opinion on the Patenting of Biotechnological Inventions Involving the Use of Biological Material of Human Origin*, October 2004, Berlin.

[28] Initially, the main issue in gene patenting was whether genetic material that had been manipulated or isolated from its natural environment and purified would be considered patentable material or a product of nature. Some opponents to gene patenting believe that no amount of manipulation is enough to label such material with the title of invention. For example, see: UNESCO International Committee on Bioethics, *Report of the IBC on Ethics, Intellectual Property and Genomics*, 10 January 2002, SHS-503/01/CIB-8/2 Rev; German National Ethics Council, ibid. However, since the US Supreme Court case of *Diamond v Chakrabarry* 447 U.S. 303, 100 S. Ct. 2204 (1980) which decided that a genetically-engineered bacterium was patentable because it was human-made and that 'anything under the sun that was made by man' was patentable, very little has not been considered patentable subject matter in the field of biotechnology and genetics. It is now settled as a matter of positive law. For more on this point, see: M. Mowzoon, *supra* Introduction, note 20, at 1082; T. Caulfield, 'Care and Innovation Agendas: The Commercialization of Genetic Research' (2003) 66 *Sask. L. Rev.* 629, at 636.

The rule for novelty and non-obviousness in genetic patenting is thus quite broad. Something can be characterised as new and non-obvious when it constitutes a real advancement and its existence was not previously documented in terms of constitution, structure (provided by genetic sequencing), process by which it is obtained, or other relevant criteria.[29]

The main contentious patenting criterion in genetics is nevertheless usefulness or utility. There is an important distinction to make between simple isolation and sequencing of genetic material and going a step further, identifying its practical application and functions. In the fast-growing field of biotechnology and genetics, scientists and institutions are racing to be the first to identify and secure exclusive (and valuable) patent rights over genetic material of interest. Since the human genome is composed of only about 30 000 genes governing millions of other biological substances and proteins, it is likely that most of these genes and gene sequences have multiple functions and interactions that will, with time and effort, gradually be discovered. When exclusive rights are granted over a whole genetic substance in exchange for some limited and incomplete information on its roles, functions, and applications, the patent-holder achieves substantial control over this material at a minor cost. In this sense, broad and vague genetic patents do not fully meet the utility criteria, lead to overcompensation and possible obstruction of research, and can therefore create some inequitable bargain between society and patent-holders. This is why things have changed gradually. We went from a trend of early patent application on genetic material of very vague and incomplete present and future implications to much more narrow and precise patent claims.[30]

---

29 O. Liivak, 'The Forgotten Originality Requirement: A Constitutional Hurdle for Gene Patents' (2005) 85:4 *Journal of Patent and Trademark Office Society* 261; R.S. Crespi, 'Patenting and Ethics – A Dubious Connection' (January 2003) *85 Journal of Patent and Trademark Office Society* 31, at 36.

30 This clearly appears from the United States Patent and Trade Office (USPTO) 2001 guidelines stipulating that any viable genetic patent claim should disclose specific, substantial, and credible utility. USPTO Utility Examination Guidelines Federal Register vol. 66 No 4, 5 January 2001 online on the USPTO website: http://www.uspto.gov/web/offices/com/sol/notices/utilexmguide.pdf (accessed 20 January 2009). These criteria have also been applied by the European Patent Office and some developing countries' patent offices; see also M. Enserink, 'Patent Office May Raise the Bar on Gene Claims' (2000) 287 *Science* 1196. Moreover, in September 2005, the US Court of Appeals for the Federal Circuit applied those guidelines in an important case on the patentability of expressed sequence tags (ESTs). In this case, the majority decided that, although the claimed ESTs were contributing to biotechnology research, they did not meet the appropriate utility requirement because the claimant did not identify the 'function for the underlying protein-encoding genes'. In *re Dane K. Fisher and Raghynath v Lalgudi,* United States Court of Appeals for the Federal Circuit, 04-1465 (Serial No. 09/619,643), 7 September 2005, Judge Rader, however,

Following this general presentation of the patent system and of some ethical and legal issues triggered by its application to genetics, it is essential to conclude this section with a few words on the territorial reach of patents, particularly on the international IP system.

3.1.2 National and International Patent Rights

Patents are territorial rights in the sense that they give the patent-holder a proprietary right over his invention within a given country. Every country can set up its own patent norms, subject to other conflicting national norms and international rules. Patents are widely enforced in industrialised countries and used more and more in developing nations under constant pressure to develop stronger IP standards. However, this is to some extent a new reality. Most of today's industrialised countries strongly resisted providing and respecting patent rights at the beginning of their economic development. Indeed, over the last century, most were focused on copying patented inventions without paying IP owners.[31] This practice ended very recently, in the 1980s, when the new industrialised countries reached a satisfactory level of social and economic growth, which put them in a position to enforce IP rights nationally. Modern developing countries have not had the same options. Even if they are far behind in terms of development, and if most believe that strong patent rights are not the best solution for their particular economic and social circumstances,[32] both developing and developed countries have to comply with the

---

enounced a dissenting opinion to the effect that the ESTs should be patentable as research tools, because they are useful for isolating and studying other molecules. There is still an ongoing debate about patents on ESTs found in a gene to determine whether they can block the use of the full patentable gene, For more on this debate, see A. K. Rai, 'Evolving Scientific Norms and Intellectual Property Rights: A Reply to Kieff' (2001) 95:2 *Northwestern University Law Review* 707.

[31] On this topic see G. Dutfield, 'Turning Knowledge into Power: Intellectual Property and the World Trade System' (2005) 59:4 *Australian Journal of International Affairs* 533, at 544–545; C. May, *A Global Political Economy of Intellectual Property Rights, The New Enclosures?* (London: Routledge, 2000) at pp. 22–44 and K. Maskus, *Intellectual Property Rights in the Global Economy* (Washington, DC: Institute for International Economics, 2000) at p. 143. More specifically, for an example of what happened in Japan, refer to C. Chien, 'Cheap Drugs at What Price to Innovation? Does Compulsory Licensing of Pharmaceuticals Hurt Innovation?' (2003) 18 *Berkeley Tech. L. J.* 853 at 863–864.

[32] A large part of most of developing countries' economies is based on imitation, something that patent law does not allow: K.A. Czub, 'Argentina's Emerging Standard of Intellectual Property Protection: A Case Study of Underlying Conflicts Between Developing Countries, TRIPS standards and the United States' (2001) 33 *Case W. Res. J. Int'l. L.* 191, at 191; K. Maskus, *Intellectual Property Rights in the Global Economy* (Washington, DC: Institute for International Economics, 2000) at p. 148.

same standards according to the agreement on Trade Related Aspects of Intellectual Property (TRIPS).[33]

Indeed, in 1995 the World Trade Organization (WTO) was created to strengthen the international trade regime. TRIPS was adopted as part of the multilateral trade agreements signed as the final act of the Uruguay Round of Multilateral Negotiations within the framework of the General Agreement on Tariffs and Trade (GATT).[34] While one role of the WTO is to deal with TRIPS implementation, enforcement, and related dispute settlement, there is another specialised international organisation established in 1970, the World Intellectual Property Organization (WIPO), that focuses on administering other existing normative intellectual property documents and on providing technical and legal assistance to countries that need it.

TRIPS' main purpose is to create an international legal structure supporting a set of minimum standards for the protection of intellectual property. All WTO member states have to comply with these legally binding principles as part of the general institutional framework set up under the WTO. This is meant to reinforce the global nature of this single intellectual property system. With its mandatory ratification and its strong and effective enforcement and compliance system, TRIPS creates a unique framework at the international level.

TRIPS supplements the 1883 *Paris Convention for the Protection of Industrial Property* and the 1886 *Berne Convention for the Protection of Literary and Artistic Works*, creating new minimum requirements relating to subject matter, scope, and enforcement of IP by all WTO states.[35] In fact, TRIPS requires all WTO member states to protect patent rights in all fields of technology for a period of 20 years from the application. This does not mean that TRIPS creates one static uniform law for everyone. Instead, TRIPS is meant to establish general minimum principles while leaving flexibility for national differentiated application, depending on specific needs and levels of development.[36] However, the effective ability of countries to take advantage (or not) of this freedom depends on various factors including external

---

[33] *Agreement on Trade-Related Aspects of Intellectual Property Rights* (TRIPS), 15 April 1994, Marrakesh Agreement Establishing the World Trade Organization, Annex 1C, Legal Instruments – Results of the Uruguay Round, vol. 31, 33 I.L.M. 81 (1994); W. Pretorius, 'TRIPS and Developing Countries: How Level is the Playing Field?' in P. Drahos and R. Mayne (eds), *Global Intellectual Property Rights, Knowledge, Access and Development* (New York: Palgrave Macmillan, 2002) p. 183.

[34] S.K. Sell, *Private Power, Public Law: Globalisation of Intellectual Property* (Cambridge: Cambridge University Press, 2003) at pp. 150–163.

[35] D.G. Richards, *supra* Introduction, note 31, at pp. 123–126.

[36] To this effect, para. 6 of TRIPS' preamble mentions *that Member States recognise: 'the special needs of the least-developed country Members in respect of maximum flexibility in the domestic implementation of laws and regulations in order to*

economic and political pressure, internal politics, local capacities, and limitations in terms of science and technology development, expertise, and infrastructures.[37] Following TRIPS' entry into force in 1995, there have been growing concerns and scepticism from many developing countries that it was not at all adapted to their needs and does not allow them enough latitude to pursue crucial public health goals.

It is in this context that the *Doha Declaration on the TRIPS Agreement and Public Health* (Doha Declaration) was proposed by a group of 80 countries led by the Africa Group, Brazil, and India. It raised considerable opposition from more affluent states but was finally adopted in November 2001.[38] The main purpose of this document is to clarify that TRIPS' dispositions should be applied in a way that allows public heath protection and encourages global access to health, especially access to affordable generic medicines for all. It also recognises that, even if intellectual property may have a positive impact on health innovation, countries need flexibility to address their domestic health needs.[39] Some believe that the Doha Declaration represents a first step in looking at TRIPS with a public interest perspective.[40] However, despite the special attention awarded to public health in the Doha Declaration, TRIPS remains an international agreement applied mainly to foster the interests of

---

enable them to create a sound and viable technological base.' Examples of such freedom can be found under sections 7 and 8 of TRIPS, which provide member states with a clear legal basis for taking measures that may diverge from generally accepted applications of the agreement by promoting *social and economic welfare, public health, nutrition, and public interest in sectors of vital importance* with the important restriction that those measures be consistent with the provisions of TRIPS itself. Moreover, art. 27(2) stipulates that states may exclude inventions from patentability in order to protect *ordre public* or morality, including to protect human, animal, or plant life or health, or avoid serious prejudice to the environment. WTO members may also exclude diagnostic, therapeutic, and surgical methods for the treatment of humans or animals (27(3)(a)) and award compulsory licences (authorising a third party to work the patent without the authorisation of the patent-holder) in limited cases and if they meet very strict criteria (art. 31).

[37] K. Balasubramaniam, 'Access to Medicine: Patents, Price and Public Policy – Consumer Perspective' in P. Drahos and R. Mayne (eds), *Global Intellectual Property Rights, Knowledge, Access and Development* (New York: Palgrave Macmillan, 2002) p. 87.

[38] Doha WTO Ministerial Declaration, 14 November 2001, WT/MIN(01)/DEC/1.

[39] This Declaration addresses issues of compulsory licences in situations of health emergency, exhaustion of rights, real and applicable differentiation in patent rules to protect public health, technology transfer, extension of the grace period for integrating TRIPS' standards for the least developed countries, etc.

[40] C.M. Correa, *Implications of the Doha Declaration on the TRIPS Agreement and Public Health*, WHO Health Economics and Drugs, EDM Series no. 12, June 2002.

intellectual property owners and promote international trade. In fact, there is growing evidence of socio-economic problems originating from TRIPS enforcement in many developing countries.[41]

3.2 SOME THEORETICAL JUSTIFICATION FOR THE INSTITUTION OF PATENTS

3.2.1 What Purpose is This System (as it Exists Today) Designed to Achieve?

This section aims to review and analyse the main theoretical arguments put forward to justify the existence of patents.[42] This is an important step in understanding the reasons put forward to justify relying on such system. Indeed, it will help us identify some of the primary objectives of the patent system, for the purpose of our further assessment.

Let us begin this subsection with a troubling quote from Edith Penrose from 1951: '[i]f national patent laws did not exist, it would be difficult to make a conclusive case for introducing them; but the fact that they do exist shifts the burden of proof and it is equally difficult to make a really conclusive case for abolishing them'.[43] This highlights what many have described as the lack of

---

41 For example, in Brazil, the enactment of a new TRIPS-compliant patent act in 1996 has had detrimental effects on availability and affordability of medicines. In fact, new patent applications have almost all been filed by non-Brazilians, medicine imports have greatly increased without similar growth in exports, and the price of drugs has increased considerably because of the lack of satisfactory anti-trust regulatory authorities. Also, in India, since TRIPS has put a stop to reverse engineering (coming up with a new process to create the same chemical entity), the domestic pharmaceutical industry has been experiencing major difficulties and has failed to secure access to important drugs for the Indian population. Moreover, price increases of 5 to 67% for patented drugs have been observed, and the associated welfare loss is being transferred to foreign stakeholders who recently recorded profits ranging between US$10 to 839 million. For more on the effects that TRIPS have had on the Brazilian and Indian economies, refer to: T.E. DeMasi and J.D. Garretson, 'PERSPECTIVE: Willful Patent Infringement Law Needs Reform' (July 28, 2003) 230 *New York Law Journal* 5 (col.1).

42 I do not aim to present an exhaustive overview of patent theory. I simply briefly introduce some of the main arguments of the major patent theories. For a deeper analysis, I encourage the reader to consult the literature on the topic, namely: J. Hughes, 'The Philosophy of Intellectual Property' (1988) 77 *Geo. L. J.* 287; B. Sherman and L. Bently, *The Making of Modern Intellectual Property Law* (Cambridge: Cambridge University Press, 1999); E. Hettinger, 'Justifying Intellectual Property' (1989) 18 *Philosophy and Public Affairs* 32.

43 E. Penrose, *The Economics of the International Patent System* (Baltimore, MD: Johns Hopkins Press, 1951) at 40 quoted in D. Vaver, 'Intellectual Property

clear grounds and agreed-upon explanation for awarding legal protection to intellectual objects, and the 'formidable task' of justifying IP.[44] For some, intellectual property is not something that can be theoretically justified because it is dependent on constantly evolving historical and cultural variables and does not have solid foundations.[45] For others, since patentable inventions are non-exclusive (they can be used by many people concurrently) and impose limits on the circulation of ideas, the burden of presenting theoretical foundations for patents falls to those who favour them.[46]

IP justification often refers to property more generally.[47] There are two main ways of rationalising institutions like property in moral philosophy: deontological and consequentialist. Deontological rationalisation refers to rights-based theories aiming at protecting what people are entitled to. It demands that decisions be made balancing the duties of some and the rights of others, determined in relation to principles that do not change according to a change in circumstances.[48] Consequentialist justification refers to the resulting positive consequences (like incentive, for example), without worrying about their underlying morality.[49] According to Nance, these two analytical tools should be used simultaneously, since '[o]ur trust in institutions like property should depend upon the existence and convergence of coherent deontological and consequentialist theories that support the rights in question and cohere with our respective views toward tangible private property and government supported private monopolies'.[50]

Today: Of Myths and Paradoxes' in P. Drahos (ed), *Intellectual Property* (Aldershot: Dartmouth Publishing, 1999) p. 485, at 495.

44 For example, see E. Hettinger, *supra* note 42, at 52; R.L. Ostergard, Jr., 'Intellectual Property: A Universal Human Right?' (1999) 21:1 *Human Rights Quarterly* 156.

45 Drahos refers to this position as post-modernist scepticism in his book: P. Drahos, *A Philosophy of Intellectual Property, supra* chapter 1, note 56, chapter 9, at p. 200.

46 B. Martin, 'Against Intellectual Property', in Drahos (ed.), *Intellectual Property*, *supra* note 5, at p. 517.

47 L. Becker, *Property Rights: Philosophic Foundations* (London: Routledge and Kegan Paul, 1977).

48 Wikipedia, the Free Encyclopedia, online: http://en.wikipedia.org/wiki/Deontological (accessed 19 May 2009).

49 H.M. Spector, 'An Outline of a Theory Justifying Intellectual and Industrial Property Rights' in P. Drahos (ed), *Intellectual Property, supra* note 5, at p. 536; J. Raz, *The Morality of Freedom* (Oxford: Clarendon Press, 1986) chapter 1; G. Davies, *Copyright and the Public Interest, Studies in Industrial Property and Copyright Law* (New York: John Wiley & Sons, 1994) at 13.

50 D.A. Nance, 'Foreword: Owning Ideas' (1990) 13 *Harvard Journal of Law & Public Policy* 757 at 767.

While the philosophical foundations for awarding patents are uncertain, it remains that theorists of different views have identified some moral and economic grounds to justify awarding exclusive protection to inventions. These can generally be categorised as follows: (1) importance of the ownership of intellectual objects in an inventor's personal development (Hegel's theory of the self); (2) respect for the inventions arising from the work of the inventor (Locke's labour theory); (3) significance of awarding exclusive proprietary rights on inventions in promoting inventive endeavours and innovation, their diffusion and commercialisation (utilitarian incentive theory); and (4) importance of patents in serving an established economic system driven by powerful agents aiming to achieve specific social and economic outcomes (Drahos' economic power theory). Each of the last three general categories[51] is addressed in the next subsections, with a special focus on the two more relevant for our analysis: the consequentialist utilitarian theory and the power theory.

3.2.2 Locke's Labour Theory

With his labour theory of proprietary rights, John Locke elaborated one of the most famous deontological justifications of the institution of private property more than 300 years ago. Locke's first principle is that everyone has property over his own person (what one decides to do with himself) and is consequently entitled to property rights over the products of his labour.[52] Hence, what is generated with the help of a person's efforts, aptitudes, and talents should be his, even if his or her labour was mixed with resources already existing in the commons.[53] Another reason for justifying property through labour is that

---

51 I left Hegel's personality theory aside since it does not apply well to patent justification as it does not seem compatible with the way the actual international intellectual property system operates. For more on Hegel's theory and on its inconsistency with the liberal theory of distributive justice I adopt in this book see G.W. Hegel, *Philosophy of Right*, trans. T.M. Knox (Oxford: Clarendon Press, 1952, lst edn, 1967 reprint) at p. 51; C. May, 'Cosmopolitan Legalism Meets Thin Community: Problems in the Global Governance of IP' (2004) *Government and Opposition* 393, at 396 et seq.

52 J. Locke, *Two Treatises of Government* (ed. Laslett) (Cambridge: Cambridge University Press, 1960) (1689).

53 J. Locke, *Second Treatise of Government* (ed. Laslett) (Cambridge: Cambridge University Press, 1988) chapter 5. It is important to highlight the difference between two labour theories: one based on natural rights to property (where property rights are owed, such as with Locke's version) and the other based on desert to property (where property rights are deserved). Under the desert labour theory, the efforts invested in labour, the risk assumed, and ethical concerns are evaluated to determine if they justify awarding property rights. This version excludes luck, intelligence and natural talents from the equation, as they are clearly not appropriate variables to assess desert. This distinction is not possible with a natural rights property theory like Locke's. For a discussion of this point, see J. Feinberg, *supra* chaper 2, note 68, at p. 16; L.C. Becker,

property plays an important role in society by encouraging people to work, something people would otherwise naturally wish to avoid.[54] Therefore, when one's labour results in valuable goods and in a society's prosperity, he should be compensated. Locke also limits the possible acquisition of proprietary rights over the product of one's labour with two provisos. The first condition is that property rights can only be awarded if there is 'enough and as good left in common for others',[55] while the second condition is that one must not get property rights on more than what he or she can use before it spoils.[56]

Although Locke was actually sceptical of the application of his theory to intellectual property, many have argued for its application to IP and patents.[57] Indeed, the production of inventions through creative effort and ideas is a type of labour that should be encouraged and patents can be essential to reward efforts and investment in research, innovation, and development.[58] However, many critiques can be formulated of Locke's theory as a philosophical foundation of patents. First, it is not always clear whether the compensation awarded by patents is proportional and justified by the efforts of the patent-holder. For example, in genetics, the fruits of one's labour will often be the result of a mix of work, highly specialised computerised research tools (for example, for sequencing and decrypting human genetic material), and a fair amount of luck. In science, moreover, it is not uncommon for several scientists to come up with the same invention almost simultaneously and totally independently.[59] It is thus hard to justify why, with the application of Locke's theory of labour, the fastest and luckiest inventors should receive all of the benefit when so many others have laboured and probably invested as much time, effort, and money in the same research endeavours.

Moreover, Locke allows property rights over anything with which one mixes his labour, suggesting that the actual labour is responsible for the quasi-total

---

Property Rights: Philosophic Foundations (Boston: Routledge, 1977) at p. 46; E. Hettinger, *supra* note 42, at p. 42.

[54] L. Becker, 'The Labour Theory of Property Acquisition' (1976) *Journal of Philosophy* 653; J. Bentham, 'The Theory of Legislation' in C.B. Macpherson (ed), *Property, Mainstream and Critical Positions* (Toronto: University of Toronto Press, 1978) at 53.

[55] This means that exclusive property rights can be awarded as long as no one is made worse off. For a discussion on that specific clause, see R.L. Ostergard, Jr., 'Intellectual Property: A Universal Human Right?' *supra* note 44.

[56] J. Locke, *supra* note 53, chapter 5, secs. 27 and 31.

[57] For example, see B.G. Damstedt, 'Limiting Locke: A Natural Law Justification for the Fair Use Doctrine' (February 2003) 112:5 *Yale Law Journal* 1179; J. Hughes, *supra* note 42, at 320.

[58] C. May, *supra* note 31.

[59] D. Vaver, *supra* note 43.

value of the fruits of labour.[60] This cumulative inventive process can create serious problems, especially in genetic research, where inventors have to build on existing knowledge and ideas, and on previous valuable inventions created and constructed over the years by many different agents within a broad social process.[61] Once the research team identifies some new, non-obvious and useful subject matter, Locke's theory allows them to obtain exclusive property rights over it and its entire market value, as if it had been developed in a social vacuum, in isolation from the broader social context.[62] In this case, awarding exclusive proprietary rights over an invention and its market value can be unfair to society and to other stakeholders, and does not demonstrate an appreciation of the importance of the numerous independent variables involved in the establishment of the actual market value of goods.[63]

Also, in aiming to reward labour, Locke seems to assume that the inventor and the patent-holder will necessarily always be the same person. This is not compatible with the patent system structure, which rewards the patent-holder with exclusive proprietary rights over some invention without worrying about the identity of the actual inventor.[64] In the fields of genetics and biotechnology, though most inventions originate from the work of some individual researchers or groups of scientists, most patents are awarded to multinational corporations and private and public research labs that employ those inventors.

Our last critique of Locke relates to his emphasis on the importance of one's ability to accomplish labour work. Locke does not talk about property redistribution, and his theory does not allow taking individuals' natural capacity differences into account. His libertarian vision brings him to focus only on capacity to generate property rights and not on compensating the less fortunate for their lack of capacity in terms of labour productivity.

Overall, Locke's vision is consistent with a libertarian theory of justice that we chose to set aside at the beginning of this book to focus on a liberal theory of distributive justice. Indeed, for us, the main concern is not the compensation for labour, but more the protection and access to health. Allowing redis-

---

60 Locke believes that things have very little importance until they are being worked on. He even proposes that 99% of objects' value is created by labour. J. Locke, *supra* note 53, para. 5, sec. 40.

61 B. Martin, *supra* note 46.

62 D.G. Richards, *supra* Introduction, note 31, chapter 2, at pp. 25–52.

63 For Hettinger, a product's market value is influenced by the productivity of competitors, the demand for the product, and the type of property institutions prevailing in a given country – all things over which the labourer does not have influence: E.C. Hettinger, *supra* note 5, at 227–230.

64 C. May, *supra* note 31, chapter 4, at 115–117. Except in the United States where the doctrine of the 'first-to-invent' still applies, so it is the first inventor that has the right over the invention.

tribution towards this end is therefore crucial to giving people more control over their lives and to target a goal of equality of opportunities.

3.2.3 The Utilitarian Justification of Property

The most common and popular justification of property is the utilitarian argument arising from the consequentialist tradition. Property theory begins with the idea that knowledge should remain freely available and unappropriated unless there is a good reason to allow its appropriation.[65]

Basically, the idea underlying this argument is that inventions are good for maximising societal benefits and that intellectual property protection is required to encourage innovative activities, production and dissemination of valuable knowledge, scientific and technological progress, and fair competition in the creation of new intellectual objects.[66] Another aspect of the utilitarian scheme is that the scarcity of ideas promotes innovation and encourages the production of more knowledge.[67] This ideology is anchored in a Western tradition that endorses a positive role for private property in economic development and does not consider inventions differently than other types of production.[68] For utilitarians, the positive effects of patents are measured in terms of their consequences on human preferences satisfaction, without taking the nature of these preferences into consideration.[69] Thus, the effects of patents on progress are positive when they play a role in improving economic development and in contributing to progress in medicine, health, agriculture, biotechnology and so on. The utilitarian justification of patents is also intended to balance their dual role, which is meant to encourage the dissemination of knowledge for long-term advancement and further development of ideas, and concurrently reward inventors with temporary exclusive proprietary rights.

---

[65] E.C. Hettinger, 'Justifying Intellectual Property', at 35–36 cited in E.R. Gold and T. Caulfield *Patents and Human Rights*, paper prepared for Justice Canada, April 2003, online on the CIPP website: http://www.cipp.mcgill.ca/data/publications/00000006.pdf (accessed 20 April 2009), at 46.

[66] In fact, as previously explained, the amount of money that needs to be invested in the development of an invention can be very high, particularly in some specialised fields of activities like biotechnology and genetics. This is one reason why many argue that incentives in the form of patents are needed to foster innovation. D.B. Resnik, 'DNA patents and scientific discovery and innovation: assessing benefits and risks' (2001) 7:1 *Science and Engineering Ethics* 29; D.G. Richards, *supra* Introduction, note 31, chapter 6, at pp. 147–151.

[67] C. May, *supra* note 1, at 127.

[68] R.L. Ostergard, Jr., *supra* note 44, at 165, D.G Richards, *supra* Introduction, note 31, chapter 2, at pp. 30–35.

[69] D. Nance, *supra* note 50, at 764–767.

There are different types of utilitarian justifications for IP relating mainly to on the one hand, its role as an incentive for innovation, dissemination, and development, and on the other hand, its role in commercialisation of inventions.

3.2.3.1 IP and innovation, dissemination, and development

The most popular utilitarian justification is to argue that IP protection creates the artificial scarcities necessary to ensure that potential inventors have sufficient financial incentive to invest in a given sector and disseminate their results.[70] As we know, when an inventor is granted a patent over an invention, he can use this right to prevent others from using the invention, recover the amount invested in developing it, disclose it to the public, and fund other research projects.[71] Proponents of this view argue that, without intellectual property rights, progress toward prevention, treatment, and cures for important health issues could be compromised or delayed.[72] Innovation and dissemination are viewed as a way to increase social welfare and inventors' reward as a mechanism to attain this goal.[73] The positive impact of patents on innovation was established in the economic literature a while ago[74] and is enshrined in the western judicial interpretation of patents' positive implications.[75] In biotechnology and genetics, this argument is said to be especially relevant because of the high costs of Research & Development (R&D) and the often lengthy

---

70 For a recent study of the negative consequences caused by the application of this justification theory (referred to in the report as the 'old IP') see Groupe international d'experts en biotechnologie, innovation et propriété intellectuelle, 'Vers une nouvelle ère de propriété intellectuelle: de la confrontation à la négociation' (2008) Montreal, Canada; K. Arrow, 'The Economics of Information: An Exposition' (1996) 23:2 *Empirica* 125.

71 S.A. Singham, 'Competition Policy and the Stimulation of Innovation: TRIPS and the Interface Between Competition and Patent Protection in the Pharmaceutical Industry' (2000) 26 *Brook. J. Int'l L.* 363, at 367–372.

72 M.F. Grady and J.I. Alexander, 'Patent Law and Rent Dissipation' (1992) 78 *Vanderbilt Law Review* 305; A.S. Oddi, 'Un-Unified Economic Theories of Patents – the Not-Quite-Holy Grail' (1996) 71 *Notre Dame L. Rev.* 267.

73 E.R. Gold, 'The Reach of Patent Law and Institutional Competence' (2003–2004) 1 *UOLTJ* 263; *Apotex Inc.* v *Wellcome Foundation Ltd.* 2002 SCC 77 (December 5, 2002), Justice Binnie, at para. 37.

74 These studies determined that patents were responsible for 15 to 25% of all innovation. For more details and reference to those studies refer to E.R. Gold et al., *supra* chapter 1, note 54, at 303.

75 For some examples on how the Courts have interpreted and justified IP, see *Fogerty v Fantasy Inc.*, 510 U.S. 517 (1994) and *Graham v John Deere Co.*, 383 U.S. 1 (1966) cited in A.K. Rai, 'Regulating Scientific Research: Rights and the Norms of Science in Biotechnology Research' (1999) 94 *Northwestern University Law Review* 77.

process of market approval.[76] Increased dissemination is meant to happen as patents are granted in exchange for the disclosure, in the patent application, of the information necessary to use and manufacture the invention. This is meant to encourage inventors to disclose what they would otherwise keep confidential.[77] However, depending on the strategy adopted by patent-holders, patents will not always play an important role in knowledge dissemination.[78]

Another argument in favour of this utilitarian justification of patent protection relates to its role as a mechanism to foster development. Indeed, in addition to increasing R&D and innovation, it is argued that patents are necessary to insure industry growth and national development through improved foreign direct investment and technology transfers.[79]

Another important aspect of the utilitarian theory of IP relates to its role in welfare maximisation. Utilitarianism requires the allocation of objects of property to those who value them the most in economic terms for the maximisation of societal benefits. In other words, utilitarianism requires the maximisation of overall welfare rather than equality in its distribution, on the basis that patents create incentives to encourage innovation. Because innovation is deemed to be good for society, allocation of patent rights should be encouraged in a utilitarian solution. Following this reasoning, property will preferably be allocated to those who can make the most productive and efficient use of it in a competitive context, allowing society to recover a maximum of benefits from those rights.[80]

---

[76] This argument is incomplete, as clearly explained in E.R. Gold and T. Caulfield, *supra* note 65.

[77] However, secrecy does not protect from independent innovations and might not always be an efficient means to protect inventions. R.S. Eisenberg, 'Patents and the Progress of Science: Exclusive Rights and Experimental Use' (1989) 56 *U. Chi. L. Rev.* 1017.

[78] For example, patent-holders could decide to disaggregate their invention into different components and claim patent rights on each of them. For more on this point, see: E.R. Gold et al., *supra* chapter 1, note 54, at 303; on the disclosure role of patents, see also *Cadbury Schweppes Inc. v FBI Foods Ltd.* (1999) 1 S.C.R. 142; F. Machlup, 'An Economic Review of the Patent System', Subcomm. on Patents, Trademarks, and Copyrights of The Senate Comm. on The Judiciary, Study No. 15, 85th Cong., 2d Sess. (GPO, 1958) at 21.

[79] E. Mansfield, 'Intellectual Property Protection, Direct Investment and Technology Transfer' *International Finance Corporation Discussion Paper No. 27*, 1995, at 11; S. Crespi, 'Models of Intellectual Property' (2002) 20 Trends in Biotechnology 451; J. Reichman, 'Universal Minimum Standards of Intellectual Property Protection under the TRIPS Component of the WTO Agreement' (1996) 29 *Int'l L.* 345.

[80] C. May, *supra* note 31, chapter 4, at pp. 122–123.

3.2.3.2 IP and commercialisation

A second important utilitarian justification for IP has to do with its role in commercialisation in terms of contribution to the manufacturing and distribution of innovations. In other words, some argue that patents are necessary for encouraging investment and coordination of the 'complex, costly and risky' commercialisation process required for taking interesting ideas and promising inventions and transforming them into useful products available in a given market.[81] This theory emphasises the importance of commercialising inventions as rapidly and efficiently as possible for the benefit of different stakeholders. For this purpose, many things must be accomplished, such as fundraising, setting up facilities to produce and manufacture the invention, establishing distribution networks, and raising public awareness about the patented product.[82] These steps come at a price, and patents are meant to exclude those who have not shared in the costs from the benefit of commercialisation. In this sense, patents can represent an important tool for securing further investment, fostering countries' competitiveness in certain areas of research, and making useful inventions available to communities.[83] Publicly recorded patents play a crucial role in helping different users of the inventions (such as developers, manufacturers, labourers, managers, investors, advertisers, and marketers) get in contact with one another and coordinate their activities around a specific invention to bring it to a stage where it can be useful to people and profitable for the patent-holder.[84] This theory is particularly adapted to the biotechnology and genetic sectors, where commercialisation costs and risks of failure are very high.[85]

---

81 Kieff articulates the commercialisation theory in F.S. Kieff, 'Coordination, Property, and Intellectual Property: an Unconventional Approach to Anticompetitive Effects and Downstream Access' (2006–2007) 56 Emory L.J. 327; F.S. Kieff 'IP Transactions: on the Theory & Practice of Commercializing Innovation' (2005) 42:3 *Hous. L. Rev.* 727; F.S. Kieff, 'Property Rights and Property Rules for Commercializing Inventions' (2001) 85 *Minn. L. Rev.* 101; F.S. Kieff, 'Perusing Property Rights in DNA' in S.F. Kieff (ed), *Perspectives on Properties of the Human Genome Project* (California: Elsevier, 2003) chapter 7, 125; for more on the commercialisation theory of patent see also G.S. Rich, 'The Relation Between Patent Practices and the Anti-Monopoly Laws' (1942) 24 *J. Pat. Off. Soc'y* 85.

82 Ibid.

83 R. Adler, 'Genome Research: Fulfilling the Public's Expectations for Knowledge and Commercialization' (1992) 257 *Science* 908.

84 H. Demsetz, 'Toward a Theory of Property Rights II: The Competition Between Private and Collective Ownership' (2002) 31 *J.Legal Stud.* 653.

85 For an enlightening and well referenced discussion on this point, refer to F.S. Kieff, 'IP Transactions: On The Theory & Practice of Commercializing Innovation' (2005), *supra* note 81.

3.2.3.3 Critique

There are problems with the way different aspects of the utilitarian justification for patents work today. In fact, they justify temporarily restricting access to the object of patents on the basis that this will encourage more production and better access to inventions in the future. This reasoning finds its origins in the idea that patents should balance both the interest of inventors in the protection of their invention, and that of society in the diffusion of new inventions. However, such balance is not easy to reach. Indeed, as Hettinger states: 'IP laws have been used more recently not as part of a social contract between creators and society, but as a tool for securing market share in an increasingly competitive global economy'.[86] For now, dissemination is constrained by the willingness and capacity to pay for accessing patented inventions. Consequently, the importance of the free flow of ideas and knowledge for societal development finds itself diluted by the application of a too strong and often unbalanced utilitarian justification of IP.

One critique of the utilitarian vision of patents arises from the fact that the encouragement of innovation is often compromised by the growing importance of patents in stimulating legal monopolies. Some suggest that biotechnology and genetic patents often do not act as incentives for socially valuable research and innovation, especially in developing countries – but more as tools used by large corporations to advance their economic agendas, gain access to more markets, and prevent other firms from penetrating specific fields of activity.[87] Indeed, the system is at the origin of inefficient 'races' between potential patent-holders who want to secure their patents first. Because only one patent ends up being granted for a single invention, this creates unnecessary duplication of research and investment in very specific, potentially profitable spheres of research.[88] Therefore, the positive impact of patent on

---

[86] E.C. Hettinger, *supra* note 42, at 50.

[87] F. Machlup, *Production and Distribution of Knowledge in the United States* (Princeton: Princeton University Press, 1962) at pp. 164–175; J.-C. St-Onge, *L'envers de la Pillule* (Montréal: Ecosociété, 2004) (on the massive production of non-innovative, 'me-too' drugs); S. Joseph, 'Pharmaceutical Corporations and Access to Drugs: The Fourth Wave of Corporate Human Rights Scrutiny' (2003) 25 *Human Rights Quarterly* 425 (on concerns about the type of innovation that patents foster); D. Noble, *America by Design* (New York: Knopf, 1982) chapter 6; R.L. Ostergard, Jr., *supra* note 44, at 165–166; M. Sakakibara and L. Branstetter, 'Do Stronger Patents Induce More Innovation? Evidence From the 1988 Japanese Patent Law Reforms' (2001) 32 *RAND Journal of Economics* 77.

[88] P. David, 'Intellectual Property Institutions and the Panda's Thumb: Patents, Copyrights, and Trade Secrets in Economic Theory and History' in M.B. Wallerstein, M.E. Mogee and R.A. Schoen (eds), *National Research Council, Global Dimensions of Intellectual Property Rights In Science and Technology* (New York: National Academy Press, 1993) 19.

innovation first established with economic studies is now being questioned. Comparative economic studies looking at different countries' patent systems and empirical research in the specific area of genetic testing suggest that there might not always be a positive link between patent rights and innovation development and access.[89] To this day, however, there is insufficient evidence to extend this conclusion to other sectors of research.[90]

Moreover, concerning the effect of IP on development, it is not clear yet from the literature, if we can establish or not a clear relationship between strong IPRs on the one hand, and foreign direct investment, foreign and local research into developing countries' diseases and technology transfer on the other hand. In other words, more empirical evidences are needed to determine if strong IPRs are, in themselves, sufficient incentive or not to attract massive foreign investment to developing countries, or to encourage technology transfer and investment in research.[91] Most development issues have to be resolved

---

89 E.R. Gold et al., 'Needed: Models of Biotechnology Intellectual Property' (August 2002) 20:8 *Trends in Biotechnology* 327; R.K. Burch, P.J.D. Smith and W.P. Wheatley, 'Divergent Incentives to Protect Intellectual Property: A Political Economy Analysis of North-South Welfare' (2000) 3 *Journal of World Intellectual Property* 169. Specifically on the effect of patents on genetic testing: J.F. Merz et al., 'Diagnostic Testing Fails the Test: The Pitfalls of Patents Are Illustrated by the Case of Haemochromatosis' (2002) 415 *Nature* 577; J.F. Merz and M.K. Cho, 'What are Gene Patents and Why are People Worried about Them?' (2005) 8:4 *Community Genetics* 203.

90 Organisation for Economic Co-operation and Development, *Genetic Inventions, Intellectual Property Rights and Licensing Practices: Evidence and Policies* (OECD, Paris: 2002), online on the website of OECD, http://www.oecd.org/dataoecd/42/21/2491084.pdf (accessed 20 January 2009), at 18 and 68.

91 Those who argue that IPRs are not determinant in development often refer to the example of the flourishing pharmaceutical industry in India, Brazil, and Argentina – nations with very weak IP norms. Most development issues must be resolved principally with the help of other international normative, economic, and political mechanisms, and many other factors need to be taken into consideration, including nations' education level, natural resources, cost of domestic labour, etc. Others, to the contrary, believe that there is a clear link between IPRs and FDI, research and technology transfer. For a more detailed analysis of the link between intellectual property rights and countries' level of development: P. Drahos, 'The Rights to Food and Health and Intellectual Property in the Era of 'Biogopolies' in S. Bottomley and D. Kinley (eds), *Commercial Law and Human Rights* (Aldershot: Ashgate Dartmouth, 2002) 227; H.E. Kettler and C. Collins, 'Using Innovative Action to Meet Global Health Needs through Existing Intellectual Property Regimes', UK Commission on Intellectual Property Rights, *Study Paper 2b*, London, 2002; C. Primo Braga and C. Fink, 'The Relationship Between Intellectual Property Rights and Foreign Direct Investment' (1998–1999) 9 *Duke J. Comp. & Int'l L.* 163, at 163; C. Correa, *Intellectual Property Rights, the WTO and Developing Countries* (London: Zed Books, 2000); Arguing for a strong link between IP and development: UK Commission on Intellectual Property Rights,

principally with the help of other international normative, economic, and political mechanisms, and many other factors need to be taken into consideration (including nations' education levels, natural resources, and cost of domestic labour). This only reveals that, before we can use this type of utilitarian argument as a reliable foundation for patent law, more empirical evidence will be needed as to the incentive function of patents when compared to other economic factors, and as to the proportionality between the means used in granting exclusive proprietary rights and the end of promoting innovative activities in genetic research.[92] As Gold and Caulfield note, the lack of economic evidence of the role of patent is symptomatic of a larger problem in patent policy: 'its reliance on *faith* and anecdotal evidence rather than on careful study and data'.[93]

In reaction to the commercialisation theory of patents, some argue that patents can interfere with commercialisation, especially in the area of fundamental research, where the patent-holders can decide to enforce their rights restrictively, potentially reducing creative activities.[94] Such an attitude toward patents can generate additional transaction costs and limit the transfer of patented products and services to the public through commercial channels. This is what has been called the *tragedy of the anti-commons*, which occurs when too many people have the right to exclude others from using resources from the commons, giving rise to their underutilisation.[95] Even if we are not in a position to conclude, one way or another with regards to the general effect of patent on preventive research at present, we can conclude that in some very specific fields – such as clinical genetics – patents appear to restrict research.[96]

Integrating Intellectual Property Rights in Development Policy, supra Introduction, note 40 at 22–26; B.M. Hoekman, K.E. Maskus and K. Saggi, *Technology Transfer to Developing Countries: Unilateral and Multilateral Policy Options* (World Bank Policy Research Working Paper 3332, 2004); H. Grabowski, 'Patent, Innovation and Access to New Pharmaceuticals' (2002) *J. of Int'l. Eco L.* 849, at 850.

92 E. Mackaay, 'Economic Incentives in Markets for Information and Innovation' (1990) 13 *Harvard Journal of Law & Public Policy* 867.

93 E.R. Gold and T. Caulfield, *supra* note 65, at 27.

94 M.A. Holman and S.R. Munzer, 'Intellectual Property Rights in Genes and Gene Fragments: A Registration Solution for Expressed Sequence Tags' (2000) 85 *Iowa L. Rev.* 735.

95 M.A. Heller, 'The Tragedy of the Anticommons: Property in the Transition from Marx to Markets' (1998) *11 Harv. L. Rev.* 621, at 624; M.A. Heller and R.S Eisenberg, *supra* note 24; G. Hardin, 'The Tragedy of the Commons' (1968) 162 *Science* 1243.

96 In fact, it is argued that innovation in drug discovery has not been seriously affected by patents on research tools, except in the area of genetic diagnostics, where some negative effects have been observed on research. Some like Kieff, sceptical about the negative effect of research tools' patents on further innovations, suggest that more empirical research should be undertaken to obtain evidence about the process of

This is why some argue that commercialisation should not occur through complete privatisation, but should also leave some space for the public domain.[97]

Most importantly, the utilitarian arguments presented in favour of IP mainly relate to the positive effects that IP can have on the competitiveness of states and companies, and on the commercialisation and availability of products in a given market. Although individuals could, in theory, benefit from IP in the long term if its incentive role were to be confirmed, the most important question this book examines is whether products emerging from innovation are truly accessible to the people who need them. The economic incentive aspect of the utilitarian theory focuses on a limited set of stakeholders, on those who value patented products the most in economic terms rather than considering the needs of all agents, including the less powerful. In fact, the only need that seems important is the need for efficiency, while the link between IP and other social needs remains unaddressed.[98] Both the incentive and the commercialisation utilitarian justifications allow us to ignore members of a community or entire nations when they do not fit into the welfare maximisation calculus. For example, utilitarianism can accept health differences between poor and wealthy individuals if the latter value some health-related products the most economically, as long as such allocation does not make some people worse off[99] and results

---

exchange failure in specific areas of research. S.F. Kieff, 'Perusing Property Rights in DNA' in S.F. Kieff (ed), *Perspectives on Properties of the Human Genome Project* (California: Elsevier, 2003) chapter 7, p. 125; J.P. Walsh, A. Arora and W.M. Cohen, 'Effects of Research Tool Patents and Licensing on Biomedical Innovation' in W.M. Cohen and A. Merrill (eds), *Patents in the Knowledge-Based Economy* (Washington, DC: The National Academies Press, 2003) p. 285, at pp. 297–305, online on the NAP website: http://books.nap.edu/books/0309086361/html/297.html#pagetop (accessed 26 January 2009); On the negative effect of patents on genetic diagnostic tools, see J.F. Merz et al., *supra* note 89; Organisation for Economic Development and Cooperation (OECD), *Genetic Inventions, Intellectual Property Rights and Licensing Practices: Evidence and Policies,* Paris, 2002, at 50.

97 J. Boyle, *The Public Domain: Enclosing the Commons of the Mind*, *supra* note 26.

98 C. May, *supra* note 1; P. Steidlmeier, 'The Moral Legitimacy of Intellectual Property Claims: American Business and Developing Country Perspectives' (1993) 12:3 *Journal of Business Ethics* 162.

99 Again, the evaluation of the notion of being worse off is problematic. As we mentioned previously, we agree with the view that being worse off can be interpreted in a relative sense, for example when something that could improve one's condition exists and this person cannot have access to it because of strong patent rights. To this effect, see: R.L. Ostergard, Jr., *supra* note 44, at 169. On the opposite view, refer to A.D. Moore 'Toward a Lockean Theory of Intellectual Property' in A.D. Moore, *Intellectual Property: Moral, Legal, and International Dilemmas* (Maryland: Rowman & Littlefield, Lanham, 1997) p. 81, at pp. 85 and 98.

in a positive impact on the overall population's health.[100] In other words, utilitarianism in health can result in awarding access to and control of genetic knowledge, products and services to those who can invest in their development and commercialisation and who can pay for them, on the grounds that their activities will likely result in overall maximisation of benefits for society (in terms of further innovation, for example).[101] However, as bluntly put by May:

> By individualising creation, by disembedding it from the social milieu from which all knowledge is drawn, IPRs deny the importance of the public realm, and by doing so reward only a small group of rights holders rather than the carriers of social knowledge, and, more importantly, ignoring the social welfare benefits of those excluded from use, not by ignorance or lack of interest, but by their poverty.[102]

Therefore, in allowing health and wealth differences, utilitarianism appears insensitive to issues of equality and distributive justice,[103] two critical concerns for equitable access to global health.

This brief overview of the utilitarian justification of IP brings us to conclude that, once again, we are not in the presence of a balanced, reliable, and complete theoretical foundation for IP. Although we are not contesting the efficiency of the IP system for serving economic and commercial purposes, it does not allow us to address our equity, access, and need issues. As it stands now, the main focus of the incentive and the commercialisation utilitarian theories remains economics, and it is imperative to balance it with other important social goals if we wish to ensure that IP can be justified in terms of distributive justice.[104]

After presenting two of the most commonly-used normative justifications of IP, we are forced to admit that none satisfactorily meets the requirements of a liberal cosmopolitan theory of distributive justice. Some arguments are

---

[100] F. Peter and T. Evans, 'Ethical Dimensions of Health Equity', *supra* chapter 1, note 6 at p. 28.

[101] This explains why, for example, only a very small fraction of all new chemicals developed between 1975 and 1996 were for the treatment of tropical diseases. For more on the effect of IP on innovation for the poor, refer to P. Cullet, 'Patent and medicines: the Relationship Between TRIPS and the Human Right to Health', (2003) 79:1 *International Affairs* 139l; J.H. Barton, 'Intellectual Property Rights and Innovation' in N. Imparato (ed), *Capital for Our Time: the Economic, Legal, and Management Challenges of Intellectual Capital* (Stanford: Hoover Institution Press, 1999) p. 123; UK Commission on Intellectual Property Rights, *supra* Introdution, note 40, at 33.

[102] C. May, *supra* note 1 at 139.

[103] B. Williams, 'A Critique of Utilitarianism' in J.J.C. Smart (ed), *Utilitarianism: For and Against* (Cambridge: Cambridge University Press, 1973) p. 75.

[104] E.C. Hettinger, *supra* note 5, at pp. 134–137.

either not adapted to the context of IP, or are based on unconfirmed assumptions about a hypothetical strong link between IP and innovation. Most importantly, none of the justifications considers the potential negative effects IP can have on those who are not in a position to own such rights.

Facing this apparent lack of a single and strong normative justification of IP, Nance proposes combining components of different theories to get a general explanation of the widespread use of, and reliance on, IP rights in today's world. [105] However, it is doubtful that any of the theories presented in the last section, taken together or separately, supply a complete justification of IP that balances deontological and consequentialist arguments in the interest of both inventors and society. To quote Nance, one should be 'sceptical of a justification of intellectual property in its present forms under any of the theories that present themselves as obvious candidates, and more sceptical of a convergence of those theories in support of intellectual property'.[106]

This leads me to the analysis of a third and last approach in our exercise of establishing some theoretical foundations for IP protection.

3.2.4 Drahos' Economic Power Theory of IP

Drahos' philosophy essentially rests on guarding society against the excess and normative risks associated with the dynamic nature and changing boundaries of IP. In an ideal world, Drahos believes in an instrumentalist approach to intellectual property rights where property is considered as a tool rather than a right. The peculiarity of Drahos' vision is that, unlike other proponents of instrumentalism, he believes that we should not focus on using IP as a tool for meeting economic ends, but instead for serving already existing moral values and distributive goals. However, Drahos' evaluation of the current IP system highlights important divergences between his ideal philosophy of IP and what he describes as an *inescapable political power theory of IP*.

In studying the development and evolution of IP law, Drahos focuses on the real importance of proprietarianism, a concept he uses to explain how IP holders are always awarded special treatment. Because property rights permit excluding and preventing others from using, selling, and producing the object of property, they allow increased private property over intellectual objects and concentration of power and sovereignty over key assets of global dependence. The effect of property rights on scarcity in knowledge essentially serves the interests of specific groups. Since IPRs are valuable assets upon which people and companies can build more possibilities, they can be defined as a form of

---

[105] D.A. Nance, *supra* note 50, at 766 and 772.
[106] Ibid., at 772.

capital, and capital is among the most important sources of power. In a way, we can say that intellectual property governs the relationship between various stakeholders, and has a direct effect on the distribution of goods and relations of dependency between IP owners and non-owners. Following this reasoning, Drahos argues that IPRs have a precarious inner logic since they are more likely to be awarded to powerful stakeholders.[107]

For Drahos, knowledge is power, and power is created by law and spread among intellectual property holders. When the law allows broadening the scope of what can be patented, society can expect that *threat power* arising from dependency relationships will end up under the control of a few. As Gold states, '[t]he purposeful omission of broader social considerations, coupled with a blind acceptance of the desirability of patents, belies a hidden libertarian agenda that favours existing distributions of wealth'.[108] In fact, obtaining ownership over intellectual objects, especially in specialised fields of activity, often requires prior scientific competence and monetary investment from patent-holders. In return, when the patent is awarded, it can generate more capital for this group of powerful agents through licensing agreements and strict control on access. For example, when lawmakers decide to extend patent protection to genes and gene-related products and services, it creates more opportunity for small elites of powerful stakeholders, who already work or may be capable of and interested in investing in genetics and biotechnology.

In this particular field of activity, we deal with universally important resources such as health knowledge and scientific progress. According to the logic of collective action, powerful agents are likely to team up in small groups to foster their common and well-defined economic interests by working on maintaining a rationalisation for the IP system.[109] In fact, the gains made by one member of the team will often benefit the rest of the group, who also share common values and interests. On the other hand, this logic does not apply in the same way to larger groups who might have a common interest in protecting the intellectual commons, such as, for example, African AIDS patients in need of cheap drugs, but for which putting up a structure and organisation to support collective action might be too much of a burden in terms of dissuasive costs. These obstacles directly influence the type of knowledge created and how it is produced and distributed.

Drahos also refers to a Rawlsian theory of justice to argue for the distribution of information instead of the excessive accumulation to which IP can give rise. Property should be treated not as the foundation of justice, but more as an

---

107 See P. Drahos, *supra* chaper 1, note 56.

108 E.R. Gold, *supra* note 13, at para. 40.

109 M. Oslon, *The Logic of Collective Action* (Cambridge: Cambridge University Press, 1965); C. May, *supra* note 31, at 22–44.

instrument for achieving well-established principles of justice; it is not a right with a fixed status, but a privilege subject to specific duties.[110] Following Drahos' distributive ideals, IPRs should therefore be used responsibly, so as to allow better access to genetics for global health improvement. Rights and property, however, form a dominant alliance in the actual IP system, and IP owners' obligations are not specified, and are often non-existent.[111] Intellectual property rights foster the interests of rights-holders without considering the associated social costs, the inequalities they can create, and the effects they can have on individuals and democratic institutions.[112] Drahos' ideal vision of IP would require the replacement of the proprietarianist view by an instrumentalist attitude supporting a different social role for IP. In fact, as Drahos mentions,

> [i]nstrumentalism would require strongly articulated conception of the public purpose and role of intellectual property. Under instrumentalism IP would be located in the context of some broader moral theory and set of values. Property rights would be morality's servants and not its drivers.[113]

One criticism of Drahos' vision is that it overemphasises the instrumental aspect of IP and treats knowledge and information as pre-existing collections from which individuals and companies *steal* for their personal benefit. Some argue that Drahos does not consider that patent-holders do not only draw from a pre-existing and static commons, but also participate in enriching it by adding to it through innovative and creative activities. This utilitarian critique of Drahos also suggests that the temporary restriction awarded by IP rights is probably the best solution to increase the bulk of information and knowledge through innovation.[114]

Drahos responds by arguing that a proprietarianist approach to IP does not necessarily encourage valuable innovation able to enrich the commons for the benefit of society as a whole. It instead contributes to creating powerful elites of property holders who participate in maintaining distributive inequalities among individuals. In fact, the IP system in its current state is often used to prevent competition and help the system's winners increase their control over more and more innovation.[115] The same legal construction that was supposed

---

110 See P. Drahos, *supra* chapter 1, note 56 at 195.

111 M. Tushnet, 'An Essay on Rights' (1984) 62 *Texas Law Review* 1363.

112 See P. Drahos, *supra* chapter 1, note 56 at 197–198.

113 See P. Drahos, *supra* chapter 1, note 56 at 224.

114 W. Van Caenegem, Book Review of *A Philosophy of Intellectual Property* by P. Drahos (1996) 8 *Bond L. R.* 217.

115 M. Goldhaber, *Reinventing Technology* (New York: Routledge and Kegan Paul, 1986) chapter 10.

to promote the diffusion of information as a common good for society is being used to restrict knowledge access by focusing almost exclusively on its economic value.[116] Because the market of ideas and knowledge is characterised by major social and economic inequalities, and since it inspires an artificial construction established to foster the interests of the more powerful property owners, the assumed utilitarian incentive theory of intellectual property seems unfounded.[117] Drahos therefore describes IPRs as special and invasive privileges that encourage power and wealth concentration in the hands of small elites, something that creates clear socio-economic and ethical struggle for the most vulnerable. Moreover, since IPRs' inner logic does not require any form of redistribution to the less powerful, Drahos argues for limited scope of IPRs without suggesting a total abolition of inventors' rewards.

The overview of the different theoretical approaches to IP presented in this section was not meant to be a comprehensive discussion of all of the different theories of property. However, it clearly demonstrates how challenging it is to try to justify property and IP. In this section, we have covered different and contradictory positions, somehow representative of the inherent contradictions present in the IP system, namely between its dual purpose of diffusion and protection.[118] Indeed, some economists in favour of maintaining an IP system (but simultaneously disturbed by some of its inbuilt and functional inconsistencies) have acknowledged that '[i]t is almost impossible to conceive of any existing social institution so faulty in so many ways. It survives only because there seems to be nothing better'.[119]

It thus appears difficult to justify patent protection with the traditional theories, particularly in the context of genetic development. For example, it is not rare in genetics to see strong public governmental participation in basic research with no strong commercial motivation; patent-holders are often private corporations and public institutions rather than actual inventors. Also, inventors are often motivated by non-commercial incentives to innovate; passion, recognition by peers, and contribution to science. Moreover, human genetic resources appear to be of very special nature, and should be considered a crucial part of the commons to be distributed following justice, need and equity considerations. In this context, Drahos' critique offers a very relevant and adapted perspective on the flaws of the existing IP system as it functions and is justified today. He addresses crucial questions regarding power

116 T.A. Lipinski and J. Britz, *supra* note 6.

117 In fact, the financial incentive for innovation is supposed to come from the market, which is flawed by major economic inequalities. B. Martin, *supra* note 46.

118 T.A. Lipinski and J. Britz, *supra* note 6 at p. 58.

119 J. Jewkes, D. Sawers and R. Stillerman, *The Sources of Invention*, 2d edn. (New York: W. W. Norton, 1969) at pp. 187–188.

imbalances created by IP and its insufficient focus on distributive justice issues.

These issues lay the foundation for my next section, where I will assess the IP system with precise benchmarks of justice.

3.3 GLOBAL DISTRIBUTION, JUSTICE AND THE PATENT SYSTEM: AN ASSESSMENT

The purpose of this section is to identify, through the lens of a global distributive justice theoretical framework, the ideal relationship that should prevail between property, genetic information, and knowledge, and assess the compatibility of the results with reality.

As explained in previous sections, although they allow patent-holders to temporarily exclude others from using, selling, or producing the object of their property right, IPRs are also supposed to encourage public disclosure of new knowledge and information in exchange for exclusive rights over it. Despite the theoretical dual role of the patent system, only one seems to prevail; it thus appears important to determine whether the system truly accomplishes its societal goal of knowledge disclosure for the common good.[120]

The extensive use of patents can give rise to important dilemmas in terms of equitable access to the object of patents, particularly when they are useful for meeting basic human needs. Indeed, although the inner logic of patents calls both for innovation protection and knowledge diffusion, it does not necessarily call for fostering equality among individuals. To examine these issues, it is relevant to test the IP system against the analytical tools developed in our theoretical framework. Our benchmark for assessing the IP system is access to genetic technologies to support health, in order to further the goal of equality of opportunities. Access is a broad concept that we will analyse through different lenses. We will begin with the evaluation of the relationship between IPRs and global access to genetic resources. We will then address the existing link between IPRs and access, in terms of availability and affordability of genetic products and services.

3.3.1 Global Access to Genetic Resources and International Intellectual Property Rights

With a global and international focus, we will first examine the compatibility of the IP system with the notion of *public good* often associated with the

---

[120] T.A. Lipinski and J. Britz, *supra* note 6, at pp. 55–56.

human genome. Following that, we will assess whether appropriation of health resources from the commons and global distributive justice in health are consistent. Finally, we will look at the possibility of equal consideration of every human being in the international IP system.

3.3.1.1 Genetic common heritage vs private appropriation of human genetic resources

The interest for genetics arises mainly from the value of emerging knowledge, which is often considered a public good. Knowledge arising from genetics and genomics can be viewed as global public goods[121] because their invention, production, and utilisation are not limited by territorial considerations (they can be used by everyone concurrently without losing value for subsequent use), and because research funding and publication in these fields are, in large part, undertaken by the public sector.[122] As things currently stand, however, the public-good nature of genetic knowledge can only benefit countries who have technology and resources to transform, apply, and use it for products development, further research, and therapeutic purposes.

The principles emerging from article 1 of the *Universal Declaration on the Human Genome and Human Rights* are a good starting point for our reflection on global access to genetics and IPRs:

> The human genome underlies the fundamental unity of all members of the human family, as well as the recognition of their inherent dignity and diversity. In a symbolic sense, it is the heritage of humanity.[123]

While this later expression has been used and supported by many,[124] its scope and practical application remain unclear. As briefly mentioned earlier, even if in fact, patenting of genetic material is possible and accepted under positive law, some people remain completely opposed to any appropriation of parts of

---

121 Which are defined as 'goods with benefits that extend to all countries, people and generations' in I. Kaul, *Providing Global Public Goods: Managing Globalization* (New York: Oxford University Press, 2003) at p. 23.

122 H. Thorsteinsdottir et al., *supra* Introduction, note 53, at 892.

123 UNESCO, *Universal Declaration on Human Rights and the Human Genome*, Paris, 1997, art. 1.

124 For example, refer to the Council of Europe Parliamentary Assembly, *Recommendation No 1425: Biotechnology and Intellectual Property,* 23 September 1999, rec. 10; HUGO, *Statement on the Principled Conduct of Genetics Research*, 21 March 1996, online: http://www.gene.ucl.ac.uk/hugo/conduct.htm (accessed 28 January 2009); Nuffield Council on Bioethics, *The Ethics of Patenting DNA: a Discussion Paper* (Nuffield Council on Bioethics: London), 2002, online on the website of the Nuffield Council: http://www.nuffieldbioethics.org/fileLibrary/pdf/theethicsofpatentingdna.pdf (accessed 30 January 2009), at 22–23.

the genome with patent rights.[125] To support their position, they argue that genetic material is very special, that it was not created by human beings, and that since we all possess copies of the human genome, it should not be possible for one to acquire exclusive property rights over what we all carry.[126] Others, on the contrary, reject the notion of common heritage and believe that property rights over the human genome are acceptable because they encourage the production of additional health and economic benefits for society.[127] A third group of thinkers rejects the concept of *heritage of humanity* for its difficult practical application, instead proposing the notion of *common resource*, which supports a form of transfer of benefits to humankind.[128] This last vision does not prevent patents over parts of the human genome if moral duties of justice and stewardship to the genome are respected.[129] In other words, the ethics and legitimacy of patents on human genetics depends on their effects on the human gene pool and on current and future generations, who all share an interest in protecting the human genome.[130] The value of the human genome for humankind is enormous. In most cases, this value is evaluated only in terms of the benefits that can be derived from resources, products, and services arising from genetic research. One fundamental challenge is thus to ensure that emerging knowledge about the human genome will benefit the entire human community.

The current IP system does not necessarily always prioritise the public domain. As previously mentioned, there was, until very recently, a strong tendency to apply very broad patentability criteria in the fields of biotechnology and genetics but things are changing gradually.[131] However, it remains, as Drahos

---

125 C. Lawson, 'Patenting Genetic Diversity, Old Rules May be Restricting the Exploration of a New Technology' (1999) 6 *Journal of Law and Medicine* 373, at 391.

126 J. Sulston, 'IP and the Human Genome' in P. Drahos and R. Mayne (eds), *Global Intellectual Property Rights, Knowledge, Access and Development* (New York: Palgrave Macmillan, 2002) p. 61; C. Joyner, 'Legal Implications of the Concept of the Common Heritage of Mankind' (January 1986) 35:1 *International and Comparative Law Quarterly* 190; see also UNESCO, *International Consultation on the outline of the Universal Declaration on the Human Genome, Summary of the response to the questionnaire*, Paris, 1997.

127 M. Spectar, 'The Fruit of The Human Genome Tree: Cautionary Tales about Technology, Investment, and the Heritage of Mankind' (2001) 23:1 *Loyola of Los Angeles International and Comparative Law Review* 1.

128 J.P. Hinojosa, *supra* Introduction, note 12; M. Kirby, 'Genomics and Democracy, A Global Challenge' (February 2003) 31:1 *UWA Law Review* 1 at 18.

129 These notions have been borrowed from the field of environmental protection. H. Rolston, *Conserving Natural Value* (New York: Columbia University Press, 1994); D.B. Resnik, *supra* note 21, at 198.

130 D.B. Resnik, ibid. at 201–202.

131 Refer to footnote 365 for a discussion and references on this point. See also J.

taught us, that proprietarianism has had a crucial role in the development and evolution of IP law by increasing the scope of private property in intellectual objects. Our actual IP system thus represents a tacit acceptance of a negative community where nobody owns the elements of the public domain, and where all states and stakeholders are free to appropriate those resources (depending on their actual economic and innovative capacities), in opposition to a positive community, where every agent automatically has joint ownership over the same elements of the commons,[132] which often results in over-consumption of the resources of the commons.[133] Those notions are directly applicable to the extensive privatisation of the human genome, made possible through the preservation of a negative community where no one opposes the privatisation of the commons, and where governments do not seem to intervene to prevent it because of its reported positive effects on innovation and related commercialisation.

In fact, even if one goal of IPRs is more knowledge diffusion for the benefit of society, the whole structure of the system of intellectual property tends to underestimate the value of the commons by 'failing to make actors and society as a whole internalise the losses caused by the extension and exercise of intellectual property rights'.[134] Those who critique the functioning of the current system believe in improved access to the human genome as a common resource. This could lead to a new egalitarian way forward in thinking about the global knowledge commons, for example, by limiting monopoly rights to encourage increased knowledge diffusion in furtherance of our goal of global distribution of genetic technologies.[135] However, some important obstacles remain, especially with regard to the growing importance, value, and protection of property rights in society, with the concomitant power relationships to which property rights often give rise, and the technical incapacity of developing countries to exploit genomics knowledge for their particular needs.[136]

The reduction of the knowledge commons for the benefit of individual appropriation is more and more pronounced[137] and, as explained by Lange, it

---

Boyle, *The Public Domain: Enclosing the Commons of the Mind*, (New Haven: Yale University Press) 2008.

132 See P. Drahos, *supra* chapter 1, note 56 at 48–49.

133 G. Hardin, 'The Tragedy of the Commons' (1968) 162 *Science* 1243.

134 J. Boyle, 'A Politics of Intellectual Property: Environmentalism for the Net?' (1997–1998) 47 *Duke L. J.* 87, at 111.

135 K. Aoki, 'Neo-colonialism, Anticommons Property and Biopiracy in the (Not-So-Brave) New World Order of International Intellectual Property Protection' (1998) 6 *Ind. J. Global Leg. Stud.* 11, at 29–35.

136 C. May, *supra* note 1 at 134–137; C. Wellman, *The Proliferation of Rights: Moral Progress or Empty Rhetoric* (Boulder: Westview Press, 1999).

137 And that, even though some steps have been taken in terms of free accessibility, for example, to the human genome draft sequence and the SNPs database. These initiatives remain the exception.

has to do with courts' and legislators' perception of the public domain as 'an unexplored abstraction instead of a field of individual rights fully as important as any of the new property rights'.[138]

3.3.1.2 Universal importance of health vs private appropriation of human genetic resources

As we saw earlier with Drahos, certain fields of activity, like genetics, have given rise to a serious concentration of power in the hands of those who have scientific and economic resources to obtain patent rights over resources upon which there is universal reliance. For example, disease-gene patenting can grant patent-holders considerable control over resources that could otherwise play an important role in improving people's health. Patentees are free to exercise their exclusive rights as they see fit for the duration of the patent,[139] without the burden of any distributive justice obligation. Therefore, even if the existence of patents is not unfair per se, the fact that property rights often take precedence over other competing rights, entitlements, and interests requires care and vigilance in their application.

Exclusive IPRs simultaneously grant economic advantage to those who have economic, knowledge, and innovative power, and often increase access costs for non-IP owners. Distribution of the costs and benefits arising from IPRs is not driven by distributive justice principles, but more by power relationships exacerbated by exclusive property rights of small elites.[140] These effects can be observed within countries, but are often more serious between countries. In fact, most people from developing nations are being left out of the IPRs system in innovative fields like genetics, as they often do not have the necessary scientific and technical power to get involved, innovate, and apply for IP protection.[141] With TRIPS, however, they must still bear the costs associated with compulsory protection of intellectual objects mostly coming from abroad. In this sense, the international system of IPRs does not ensure that people from developing countries have their health needs satisfied; in some cases, it even contributes to engendering health gaps within and between nations.

---

138 D. Lange, 'Recognising the Public Domain' (1981) 44:4 *Law and Contemporary Problems* 147, at 178.

139 That is, subject to other existing national and international legislation.

140 J.W. Singer, *Entitlement: The Paradoxes of Property* (New Haven: Yale University Press, 2000), at p. 68.

141 We need to nuance here because over the last few years we have witnessed progress in genomics in a few developing countries, like Thailand, who have succeeded in research relating to the need of their populations, see for example B. Séguin et al. 'Human Genomic Variation Initiatives in Emerging Economies and Developing Countries', *supra* Introduction, note 30.

The principles established in the first part for global equitable distribution of genetic benefits recognise the universal importance of health for every human being, no matter where they live. My theoretical analysis has led me to conclude that since health is a crucial human need, securing it is necessary to avoid serious harm and to develop a normal range of opportunities. This is what inspired my argument in favour of genetic-benefit distribution in an effort to compensate for global health inequalities and deviations from normal functioning, two important elements for equality of opportunity. This global distributive justice framework for equitable access to health should encourage the establishment of international principles and institutions. However, these values are not TRIPS' primary focus.

3.3.1.3 Universal consideration of every human being in the international IP system

We have to address another important aspect of our theoretical framework in relation to IP: the cosmopolitan focus of our approach to justice. To this end, I will assess the global structure supporting the international IP system to determine whether it conforms to the principle of universal consideration of every human being required by moral cosmopolitanism. To help me in this task, I will determine whether *the international IP system* can qualify as a just global basic structure and whether it responds to shared human interests in equality of opportunity.

Does the international IP system correspond to a just global basic structure? Some who are opposed to a global application of principles of distributive justice argue that the existence of a coercive network of law is absolutely essential to engage in redistribution, and that this type of network does not exist on the global scene.[142] This argument does not stand, especially when applied to the international IP system, which easily qualifies as a coercive legal network.[143]

In my theoretical framework, I supported a distribution of benefits arising from the commercialisation of a common resource (the human genome) extended to the global scene, arguing that boundaries and citizenship should not limit the scope of social cooperation, loyalty, and obligations. I therefore adopted a cosmopolitan focus referring to each individual as a unit of moral concern, and concluded that everyone affected by institutional distributive

---

142 For example, refer to M. Blake, *supra* chapter 1, note 53, at 291–292.

143 Indeed, although its main goal is not to redistribute resources, the international IP system requires every WTO member state to adhere to and comply with TRIPS, which is complemented by a unique and effective enforcement mechanism.

arrangements should be given the chance to secure a normal range of opportunities for himself, in order to have access to a *decent* life.[144] All that cosmopolitanism requires is a vision of sovereignty that does not constitute an arbitrary limit on the scope of justice.[145]

Indeed, principles of distributive justice can continue to impose obligations for the satisfaction of individual rights both on states (which, for now, remain the primary agents of distributive justice), and on other institutional actors (who may also be viewed as potential agents of international justice obligations). A significant challenge is to find a balance between the order supported by sovereignty and the pursuit of justice through moral universalism. Some propose that state and non-state agents should work toward an agreement on general ethical principles, values, and duties that international society should internalise and promote as a group. It would extend the frontiers of communities, despite the different cultural and community allegiances.[146] In fact, these principles would serve as a basis for the establishment of a broad political community outside the boundaries of states. It is relevant to mention that we can actually observe a proliferation of transnational advocacy networks driven by common values and aiming to reconstruct the scope and limits of state sovereignty on the international scene by denouncing inequities.[147]

Therefore, even if there is, to this day, an institutional inability to implement principles of distributive justice at the international level, these principles have an important role to play in identifying future courses of action to develop institutional capacity and governance in an emerging global society.[148] This could result in the expansion of democracy beyond the state structure and help challenge the mechanisms of non-democratic globalisation from above which are supported by the International Monetary Fund, the World Bank, and the World Trade Organization.[149]

---

144 I refer the reader to the theory chapter for a comprehensive discussion on cosmopolitanism.

145 O. O'Neill, 'Justice and Boundaries', *supra* chapter 1, note 58 at 69.

146 A. Linklater, *Transformation of Political Community* (London: Polity Press, 1998) at p. 167; R.A. Payne and N.H. Samhat, *Democratizing Global Politics, Discourse Norms, International Regimes and Political Community* (New York: State University of New York Press, 2004); J. Thompson, 'Community, Identity and World Citizenship', in D. Archibugi, D. Held and M. Köhler, *Re-imagining Political Community: Studies in Cosmopolitan Democracy* (Stanford, CA: Stanford University Press, 1998) p. 179, at p. 191.

147 M.E. Keck and K. Sikkink, *Activists Beyond Borders: Advocacy Networks in International Politics* (Ithaca: Cornell University Press, 1998) chapter 6.

148 C.R. Beitz, *Political Theory and International Relations, supra* chaper 1, note 1 at p. 271; A. Buchanan, *supra* Introduction, note 30, chapter 4, at pp. 191–230.

149 R. Falk, *On Humane Governance: Toward a New Global Politics* (University Park, PA: Pennsylvania State University Press, 1995) at 125.

This brings us to the notion of global basic structure, a concept referring to the existing, well-established global scheme, characterised by international legal systems mostly controlled by private property and trade regimes.[150] Some, like Beitz and Barry, argue that this global basic structure creates a *pattern of global interdependence* between participating states, and that this system implies mutual cooperation required by global distributive justice.[151] Drahos is more sceptical; he does not automatically link states' economic interdependency and their involvement in a scheme of mutual cooperation. He believes that the global structure is very heterogeneous in terms of group beliefs, moral codes, and cultural practices, and that it is therefore difficult to identify global principles of justice for mutual cooperation. He does not rule out global distributive justice theory, but decides not to pursue it because of the problems he anticipates with its application.[152]

Whatever qualification we apply to the existing global structure, whether we believe it is a site of interdependence or of mutual cooperation, there is no reason why it should not also be a subject of justice.[153] Unfortunately, justice, as we envision it, does not seem to direct the existing global structure. Indeed, for some, the global community emerging from this structure mirrors the economic inequalities and gives rise to the same concentration of power observed at state levels. This makes this international structure unsuitable to further global community interests.[154] Other scholars, including Cox and Richards, believe that it is not a few states which dominate the world order, but more a dominant ideology with a central mode of production and distribution infiltrating every state.[155] This new elite occupies a privileged space on the international scene, without representing individual interests and without

---

[150] The basic global structure is composed of regional and international economic agreements like NAFTA and TRIPS, international monetary schemes like The World Bank and the International Monetary Fund, and international human rights standards.

[151] C.R. Beitz, *Political Theory and International Relations*, *supra* chapter 1, note 1 at p. 145; B. Barry, *The Liberal Theory of Justice: Critical Examination of the Principal Doctrines In A Theory of Justice by John Rawls* (Oxford: Clarendon Press, 1975) at 129.

[152] See P. Drahos, *supra* chapter 1, note 56, chapter 8, at pp. 170–198. I have already analysed similar critiques of global distributive justice in the theory chapter, basing my argument on the special character and universality of health as something that transcends countries' cultural differences, special beliefs, and moral codes. For that reason, I decided to reject a communitarianist approach to justice.

[153] A. Buchanan, *supra* Introduction, note 30, chapter 2, at pp. 73–74.

[154] For example, see May on this point: C. May, 'Cosmopolitan Legalism Meets Thin Community: Problems in the Global Governance of IP', (2004) *Government and Opposition* 393 at 410.

[155] R. Cox, *Approaches to World Order* (Cambridge: Cambridge University Press, 1996).

being democratically accountable. As explained by Hymer, the power of multinational corporations as principal agents of globalisation is replacing the traditional authority of the state: '[w]hen a corporation invests abroad it not only sends capital and management out but also establishes a system for drawing foreign capital and labor into an integrated world network'.[156]

One issue that is not debated is that the shared global order currently in place is established by the most fortunate and imposed on the worst-off. As it represents an important part of the global order, I here subject the international IP system to assessment from a global distributive justice perspective. It should contain mechanisms to ensure that everyone has access to a certain level of health (influenced by access to genetic knowledge and products), in order to be able to secure a normal range of opportunities. This duty of justice in health originates from different theoretical sources already discussed in the first part of this book. Regardless of which theoretical reason we choose to justify justice obligations (we argued for a duty to avoid harm in our theoretical framework), the economic disparity between different countries and regions of the world makes it impossible to address issues of distributive justice and property rights over the human genome without polarising the world into two large groups: developed and developing countries.[157] In fact, even if we all share our genetic background with everybody else on the planet, the technology for unlocking the value of these human genetic resources is patented mostly in developed countries and sometimes in developing countries, for the most part by stakeholders from the developed world. This factor, among many others, can influence access to genetic benefits by individuals of different regions of the world.[158]

This dichotomy is illustrated in the application of TRIPS, which was created primarily to secure benefits for the IP owners, encourage international trade, and establish a system that fosters minimal standards of IP for every WTO member state. In fact, the global scheme broadens patentability criteria for a minimum 20-year period and has a general negative effect on developing countries, and people from those countries, in terms of access to health and genetics. The basic mission of the WTO is to foster a uniform system of liber-

---

[156] S. Hymer, *The Multinational Corporation: a Radical Approach* (Cambridge: Cambridge University Press, 1979) at p. 76; D.G Richards, *supra* Introduction, note 31, chapter 4, at pp. 79–111. See also A. Clapham, *Human Rights of Non-State Actors* (New York: Oxford University Press, 2006).

[157] P. Kameri-Mbote, *Property Rights and Biodiversity Management in Kenya* (Act Press: Kenya, 2002).

[158] For statistics on this point, refer to World Bank, World Development Indicators 2005, Geneva, 2005, online on the website of the WB, http://devdata.worldbank.org/wdi2005/index2.htm (accessed 2 February 2009), table 5.12.

alised trade at the global level.[159] In this sense, TRIPS is a tool to encourage international capitalism and strengthen existing global inequalities. This obviously creates tensions between exporters and importers of IP goods. Some scholars and numerous reports argue that, in prioritising protection instead of diffusion, this system mainly targets the needs of wealthy inventors and IP owners, and results in shrinking the bulk of public knowledge. They also believe that the global standardisation and proliferation of IP norms fostered by TRIPS does not encourage more and better knowledge diffusion and dissemination as a single patent application in one country only is sufficient to ensure diffusion.[160] TRIPS can thus harm poorer producers and the public at large by not taking egalitarian grounds into consideration and not fostering public welfare goals.[161] The proponents of this position highlight that IP globalisation can be very costly for poor countries and have disastrous effects on global welfare with very little benefit for the majority in return.[162] However, as we know, others argue instead that increasing IP protection globally can result in positive welfare effects for developing countries by encouraging innovation, foreign direct investment, and technology transfer.[163]

Regardless of the view we adopt on the practical short- and long-term effects of patents on innovation in developed and developing countries the fact that TRIPS emerged from negotiations undertaken under the auspices of the WTO explains most of its effects on global access and distribution of health and genetics. Some argue that since the *Uruguay Round Agreements* (including the *Multilateral Agreements on Trade in Goods*, the *General Agreement on Trade in Services*, and TRIPS) were voluntarily signed by various countries – both from the developed and the developing world – it

159 As very well explained by D. Moellendorf, 'The World Trade Organization and Egalitarian Justice' (January 2005) 36:1/2 *Methaphilosophy* 145, at 150–152.

160 P. McCalman, 'Reaping What You Sow: An Empirical Analysis of International Patent Harmonization' (2001) 55 *Journal of International Economics* 161.

161 And that, even though some dispositions aim to find some balance between innovation and diffusion with exceptions to patent protection. This will be discussed further in the last section of this chapter.

162 For example, refer to: UNCTAD, *The TRIPS Agreement and Developing Countries,* UNCTAD, Geneva, 1996; UNDP, *Human Development Report 2001*, Geneva, 2001; UK Commission on Intellectual Property Rights, *Integrating Intellectual Property Rights in Development Policy, supra* Introduction, note 40, at 5; K. Maskus, 'Normative Concerns in the International Protection of Intellectual Property Rights' (1991) 14 *World Economy* 403.

163 For example, refer to E. Su, 'The Winners and the Losers: The Agreement on Trade-Related Aspects of Intellectual Property Rights and Its Effects on Developing Countries' (2000) 23 *Hous. J. Int'l L.* 169; Gold et al., *supra* chapter 1, note 54, at 328.

must have been satisfactory and beneficial to everyone.[164] This reasoning implies that those agreements leave everybody in a better position and that consent is sufficient to infer legitimacy. However, what the proponents of this view seem to ignore is that TRIPS' negotiations occurred within a global scheme and under an international organisation characterised by numerous power inequalities.[165] As explained by Buchanan, 'unless the background institutions of the basic structure are just, injustices may be perpetuated by voluntary agreements'.[166] In this case, developing countries were made to realise that they did not really have any other choice but to accept TRIPS' conditions negotiated in a context of economic oppression and power imbalance.

Consequently, in reply to the question of whether the international IP system qualifies as a just global basic structure, we can say that TRIPS is part of a global scheme imposed by the most affluent on the less fortunate, a structure that does not qualify as just when assessed from our global distributive justice perspective. Indeed, global distribution of health benefits is a crucial element of our ideal theory of justice, and it does not seem to be central to TRIPS, which was designed by a few stakeholders to further their own private interests in a global structure they run. Undeniably, the international political system and, more specifically, TRIPS, as they work now, give rise to power struggles that almost always take our attention away from the universal principles and values standing at the basis of a cosmopolitan vision of humanity.

Does the international IP system offer a response to shared human interests in equality of opportunities? Our normative cosmopolitan principles require that individuals be considered the 'normative epicentre of a system of functionally plural sovereignty'.[167] Some could argue that since the international IP system was established in an agreement between sovereign states for minimum standards of protection for both inventors and the public, it is therefore compatible with a cosmopolitan ideal, since it considers everyone. An analy-

---

[164] Following a complex process of negotiations, developing countries agreed to TRIPS in exchange for concessions in other trade-related sectors like agriculture and textiles. *Marrakesh Agreement Establishing the World Trade Organization,* Annexes 1A, B and C, 33 I.L.R. 1197 (1993).

[165] For an enlightening discussion on this point, refer to K. Raustiala, 'Compliance and effectiveness in international regulatory cooperation' (Summer 2000) 32:3 *Case Western Reserve Journal of International Law* 387; P. Gerhart, 'Reflections: beyond compliance theory – TRIPS as a substantive issue' (Summer 2000) 32:3 *Case Western Reserve Journal of International Law* 357, at 371.

[166] A. Buchanan, 'Rawls's law of peoples: Rules for a Vanished Westphalian World' *supra* Introduction, note 49, at 705.

[167] A. Kupler, *supra* chapter 1, note 5.

sis of the historical and political context of TRIPS' adoption, however, leads us to conclude that its creation, adoption, and enforcement have not been driven by shared morality but by the interests of the more powerful.[168] Overall, in increasing the scope and reach of intellectual property at the international level, TRIPS strengthened the property power of the most affluent stakeholders of the world.[169] As May describes, 'the legal rules encapsulated within the TRIPS represent the triumph of the knowledge structure's agenda of the metaphorical links between knowledge and property'.[170]

This clearly demonstrates that TRIPS did not result from negotiations undertaken in conditions of voluntary mutual assistance. Each state that had a minimum amount of power took care of advancing its own interests and advantages as much as it could.[171] Moreover, the WTO is an organisation which has no democratic features and is subject to the same economic inequalities we find at the state level, something which greatly limits democracy at the international level because of the lack of central authority.[172] In a way, it is 'extra-governmental' and beyond the direct reach of the electorate.

In response to this situation, some argue for increased decentralisation and prioritisation of national interests, both on the international and the national scenes, instead of trying to agree on global shared interests. One argument to support this view is that different cultures call for different actions, and attempts by some states to establish international moral standards could lead to moral imperialism.[173] There are a few problems with this national interest view. One is that, in reality, the national interest is often the expression of the interests of small number of elites who have enough power to neutralise the interests of other groups. This is clearly illustrated by the outcome of TRIPS' negotiations. The strong multinational lobby of the pharmaceutical industry, represented in different industrialised countries, vigorously promoted its own

---

168 For a detailed analysis of TRIPS' context of adoption and operation, I suggest that the reader refer to S.K. Sell, *supra* note 34. The author even questions the legitimacy of the agreement because of its coercive negotiation context and absence of mutual benefits.

169 C. Arup, 'Competition over Competition Policy for International Trade and Intellectual Property' (1998) 16:3 *Prometheus* 367, at 376.

170 C. May, *supra* note 31, at p. 34.

171 For Buchanan, 'the state is nothing more than a discretionary association for the mutual advantage of its citizens', A. Buchanan, 'The Internal Legitimacy of Humanitarian Intervention' (1999) 7 *Journal of Political Philosophy* 71.

172 J.S. Dryzek, 'Transnational Democracy' (March 1999) 7:1 *Journal of Political Philosophy* 30, at 33.

173 H.J. Morgenthau, *Politics Among Nations: The Struggle for Power and Peace*, 6th edn (New York: Alfred A. Knopf, 1985).

interest in a universal minimal coverage of patent protection.[174] In so doing, these multinational corporations had a large impact on the adoption of the TRIPS agreement as it now stands.[175] They were very efficient in convincing state policymakers to promote their interests. By letting the pharmaceutical lobby implicitly lead the negotiations and further their economic agenda in the name of whole nations, developed countries like the US and Japan, as well as the EU, agreed to translate private interests into matters of public interest. This concurrently left out the concerns of other national interest groups like patient groups, NGOs, and citizens.[176]

Another difficulty arising from a focus on national interest is that it assumes that state officials will necessarily act to promote their populations' interests. It thereby ignores the fact that many poor countries are very poorly governed. In fact, many of them are run by corrupt officials, not always democratically elected and often more concerned about advancing their own interests than those of their populations. It is thus fair to say that individuals forming the populations of such unstable states cannot count on their governments to represent them adequately, and would therefore derive great advantage from the application of global principles of shared morality. Moreover, we cannot ignore the global institutional context's role in maintaining some of these corrupt, undemocratic, and unstable governments. As long as affluent states recognise the effective political and trade power of unstable governments and will design global normative instruments like the international intellectual property scheme hand-in-hand with them, global inequities and poverty will persist.[177]

The liberal cosmopolitan distributive justice theory adopted in this book calls for a shared morality between all societies and humans to meet basic and universal needs and values. Stuart Hampshire defines basic ethical principles as 'those that if followed, help avert the worst harms to which all human

---

174 In fact, large pharmaceutical companies believe that patent protection is the most important tool to uphold their investment in R&D and innovation, in addition to furthering their corporate strategies as explained in F.M. Scherer, 'Le Système de Brevet et l'Innovation dans le Secteur Pharmaceutique/ The Patent System and Innovation in Pharmaceuticals' (2000) 1 *Revue Internationale de Droit Economique* 110, at 112.

175 M. Ryan, *Knowledge Diplomacy: Global Competition and the Politics of Intellectual Property* (Washington DC, Brookings Institution, 1998) at pp. 67–72; S.K. Sell, *supra* note 34, at pp. 23–25.

176 Many interest groups who stand on diametrically opposed grounds to pharmas exist everywhere in the world and are increasingly influencial on the global and national scenes.

177 T.W. Pogge, 'Symposium on World Poverty and Human Rights', *supra* chapter 2, note 97, at 7.

beings are vulnerable, those principles to which adherence is necessary for people being able to lead decent human lives'.[178] As I argued in the first two chapters, it is important to identify spheres of common and universal interest that can forge a sense of global identity necessary to find a consensus on common ethical principles relating to universal concerns for the well-being of every individual.

In the broad area of international law, this shared morality should be served by integrating distributive justice standards in trade relations, labour and environmental law, and in the global intellectual property rights scheme for a more equitable international allocation of the health benefits arising from biotechnology and genetics. For now, however, the social utility function of IP is interpreted quite narrowly as we witness property rights owned by corporate stakeholders on products which make use of elements of the public commons. Unfortunately, the international IP system can be viewed as a foil designed to benefit the owners and managers of multinational capital invested in funding, creating, and supplying the knowledge-based inventions in different markets worldwide.[179] TRIPS has the status of public international law and functions mostly without having to take health and welfare needs of the world's poor majorities into great consideration. TRIPS contains some exceptions and flexibility, but these are often given minimum consideration in practice as I will discuss in more length under section 3.3.2.2. In this sense, TRIPS is compatible with the neo-liberal ideology that supports a fundamental civil right of *freedom of trade* for every individual.[180] In awarding precedence to property rights and freedom of trade, this view concurrently legitimises inequalities in health and differences in opportunities.

This tendency needs to be corrected at the normative level if we want to hope for global distributive justice and social welfare. Actions have already been taken in this direction as a result, among other actions, of NGO campaigns denouncing the effects of TRIPS on access to essential drugs. Indeed, more and more, interesting innovative models are being put into place like public private partnerships that aim to develop vaccines and drugs to fight aids consortiums and patent pools established between industries, universities, private foundations, NGOs and governments to encourage collaborative initiatives to improve

---

[178] S. Hampshire, *Innocence and Experience* (Cambridge, MA: Harvard University Press, 1989) at p. 90; see also A. Linklater, *supra* note 146, at pp. 167–170.

[179] R.A. Payne and N.H. Samhat, *supra* note 146, chapter 5, at pp. 135–140.

[180] E.-U. Petersmann, 'National Constitutions and International Economic Law' in M. Hilf and E.-U. Petersmann (eds), *National Constitutions and International Economic Law* (Deventer: Kluwer, 1993) p. 3; E.-U. Petersmann, 'The WTO Constitution and Human Rights' (2000) 3:1 *Journal of International Economic Law* 19, at 21–23.

innovation towards global health.[181] Yet, much more needs to be done for the global IP system to mirror a shared human interest in health and to promote equality of opportunity through distributive justice mechanisms.[182]

Having assessed the international dimension of the IPR system with different aspects of global access to genetic resources, we will now evaluate the compatibility of IPRs with other facets of access: availability and affordability of genetic research tools, products, and services as mechanisms to further equality of opportunities.

3.3.2 Patents and Access in Terms of Availability and Affordability

The philosophy of article 19 of the *Universal Declaration on the Human Genome and Human Rights* is a good starting point for our reflection on access to genetics in terms of availability and affordability:

> a) In the framework of international co-operation with developing countries, States should seek to encourage measures enabling:
> i) assessment of the risks and benefits pertaining to research on the human genome to be carried out and abuse to be prevented;
> ii) the capacity of developing countries to carry out research on human biology and genetics, taking into consideration their specific problems, to be developed and strengthened;
> iii) developing countries to benefit from the achievements of scientific and technological research so that their use in favour of economic and social progress can be to the benefit of all;
> iv) the free exchange of scientific knowledge and information in the areas of biology, genetics and medicine to be promoted.
> b) Relevant international organizations should support and promote the initiatives taken by States for the above mentioned purposes.

The main goal of this section is to determine how the need for access to genetics is realised in relation to availability and affordability, and whether the actual IP scheme is designed to meet and prioritise this need.

---

[181] T. Bubela et al., *Respecting, Promoting, and Protecting Traditional Knowledge: A Comparative Case Study of Brazil, Kenya, and Northern Canada*, 2008, online www.theinnovationpartnership.org (accessed 20 May 2009); D. Castle and E.R. Gold, 'Traditional knowledge and benefit sharing: from compensation to transaction' in P. Philips and C. Onwueke (eds), *Accessing and Sharing the Benefits of the Genomics Revolution* (Dordrecht: Springer, 2006).

[182] Groupe international d'experts en biotechnologie, innovation et propriété intellectuelle, *supra* note 70, at p. 29.

3.3.2.1 Spheres of genetic access affected by patents

One particularity of genetics is that access to what has already been discovered is necessary to push the science further and discover alternative applications. As Sulston notes, 'it is not possible to reinvent a human gene',[183] and since the most important gene applications are often discovered following many years of cumulative research, access to prior work is of critical importance for the scientific survival of the whole field. Property rights in intellectual objects can have an impact on different spheres of genetic access. For example, patents can influence availability of research tools in placing a temporary embargo on crucial elements necessary for the advancement of genetic research. Patents can also affect the availability of genetic tests and services in vulnerable communities and populations as they can influence affordability of products and services.

Availability of research tools In genetics, research tools are the input needed to develop, discover, or invent innovative heath-related products. For example, these tools include DNA, genes, sequencing techniques, genetic bio banks, stem cells, cell lines, single nucleotide polymorphisms (SNPs), genetic knowledge, etc.[184] In the course of their research projects, investigators will often need to access, analyse, and duplicate many research tools. When they are patented, their use will increase the cost of research if, for example, patent-holders impose high licensing fees, or if the use of many patented tools is needed for the same project and some patent holders are reluctant to license their rights. This latter concern is especially relevant in genetic research, where it is not uncommon to see multiple (patented) genes and gene sequences involved in the expression of one single disease. As discussed earlier, this has been referred to as the *tragedy of the anti-commons*, occurring when too many holders of property rights are in positions to exclude others from using resources from the commons.[185] Some, like Kieff, respond to such concern arguing that patents on research tools will instead encourage their commercialisation and accessibility and that many patent-holders will naturally be inclined to widely license their right in useful research tools in order to

---

[183] J. Sulston, *supra* note 126, at p. 86.

[184] WHO, *Genetics, Genomics and the Patenting of DNA: Review of Potential Implications for Health in Developing Countries*, *supra* Introduction, note 21, at p. 39.

[185] This can happen with the ongoing race taking place in genetics and genomics. Various companies have taken out patents on multiple DNA sequences in a race to control entire genes, and they could use their rights in these sequences to prevent other scientists from using or studying them. Indeed, they could refuse to license their rights or charge high prices for their use. For example, M.A. Heller, *supra* note 95, at p. 624; D.G. Richards, *supra* Introduction, note 31, chapter 2, at pp. 36–38.

become more famous and receive academic recognition from their peers.[186] As briefly mentioned earlier, there is not enough data for now to conclude that patents are at the origin of a *tragedy of the anti-commons*, but there are some indications that patents, at least in certain cases, can negatively impact the conduct of genetic research and inhibit science.[187] This is especially true for research undertaken by research bodies operating with limited capital and interested in areas of research that are likely to be less profitable. Because of high licence costs, some might not be able to use patented, expensive tools to progress in their research and to further their efforts towards commercialisation, in contrast to those who are able to afford these instruments to further their own research and commercialisation agendas, mostly in profitable research areas. As bluntly put by Drahos and Mayne, '[i]f the poor want more patent based R&D for malaria they will have to hope that is overtakes obesity and impotence as a problem in western societies'.[188] In the same vein, some even go as far as to argue that market forces and property rights contribute to establish new standards of health, normality, and disease.[189]

Availability of genetics products and services Another availability issue concerns the development and distribution of genetic tests and services to people living in countries representing non-lucrative markets. As just mentioned, patents on research tools can, in some cases, slow progress in areas of special relevance to developing countries, which therefore have to rely on what is being produced and patented in more affluent countries.[190] Patent protection might not always encourage availability of products and services in developed and developing countries. In fact, as already mentioned, there is an

---

[186] For more on this argument, see: F.S. Kieff, 'Facilitating Scientific Research: Intellectual Property Rights and the Norms of Science – A response to Rai and Eisenberg', *supra* note 30.

[187] Organisation for Economic Development and Cooperation (OECD), *Genetic Inventions, Intellectual Property Rights and Licensing Practices: Evidence and Policies,* Paris, 2002, at 50.

[188] P. Drahos and R. Mayne (eds), *Global Intellectual Property Rights, Knowledge, Access and Development* (New York: Palgrave Macmillan, 2002) Introduction, at p. 7. Some countries have taken matters in their own hands, as we discussed earlier, but it is still a minority as most developing countries still have to rely on what is being accomplished in the developed world, see B. Séguin et al., 'Human Genomic Variation Initiatives in Emerging Economies and Developing Countries', *supra* Introduction, note 30.

[189] For a detailed explanation of this theoretical view, see M. Martone 'The Ethics of the Economics of Patenting the Human Genome' (1998) 17 *J. Bus. Ethics* 1679, at 1679.

[190] Developing countries are also often technologically unable to develop tests, products, and services, as we will see in the last section of this chapter.

ongoing debate as to whether patents encourage innovation or not, and there is a lack of empirical evidence to support any position. One exception to this might be the field of diagnostic genetic tests, where the prospect of patent rights has not been an incentive to development.[191] In fact, there is evidence that patents have encouraged early release of genetic products that might not have been of the best quality and reliability.[192] Moreover, depending on the scope of the patent granted, patent-holders can control how the product will be used in clinical and research settings, its cost, and the mode of analysis to use with the product.[193] This could greatly influence availability of health-related products, especially to the most vulnerable populations and individuals. In response to such important concerns, in February 2006, members of the OECD adopted guidelines for governing the licensing of genetic inventions used in health care settings. Those guidelines are meant to encourage both innovation in genetics and fair economic returns, rapid dissemination, and access to diagnostic and therapeutic products and services.[194]

Another factor influencing availability, and which is not patent-related, is the lack of infrastructure within developing nations. Genetic compounds, products, and services will not always be patented in developing countries because of the absence of a market and the lack of possible financial return and profits. Theoretically, this would mean that those countries could use the technology, products, and services without restriction. However, in these cases, availability of genetic services is not influenced by patent rights, but instead by a country's research, medical, and manufacturing infrastructure, and by its lack of trained professionals. As we will briefly explain in the next section, it is important to realise that although patents might have a role to play in availability issues in developing countries, it is only one issue to consider and must not be blamed for everything. It is thus crucial to highlight the importance of developing appropriate infrastructures and training programmes for availability of genetics in these countries.

---

[191] In fact, the following article reports evidence that the development and the use of many genetic tests were undertaken by laboratories other than those planning to apply for patent protection: J.F. Merz et al., 'Diagnostic Testing Fails the Test: The Pitfalls of Patents are Illustrated by the Case of Haemochromatosis', *supra* note 89.

[192] T.A. Caulfield and E.R. Gold, 'Genetic Testing, Ethical Concerns, and the Role of Patent Law' (2000) 57 *Clinical Genetics* 370; E.R. Gold, 'From Theory to Practice: Health Care and the Patent System' (September 2003) *Health Law Journal*, Special Edition 2003: Precedent and Innovation: Health Law in the 21st Century.

[193] Subject to market forces, competition and existing legal norms.

[194] OECD Council, *Recommendation on the Licensing of Genetic Inventions*, 23 February 2006. C (2005)149/Rev1, online on the OECD website: http://www.oecd.org/dataoecd/39/38/36198812.pdf (accessed 2 February 2009).

Affordability of genetics products and services Affordability of products and services is another important sphere of access in genetics. Patents awarded in developed countries can have a direct effect on the price of genetic products and services both in developed and developing nations. In fact, since most developing countries do not have the necessary infrastructure to develop and manufacture health-related products, they have to rely on more affluent countries for the supplies they need.[195] Thus, the effect of patents on the cost of genetic technology and services will also be transferred to importing, developing countries. We know that awarding patent rights over intellectual objects allows patent-holders to license their rights under conditions they set or sell patented objects at a price they unilaterally fix.[196] This is meant to help patent-holders recoup the capital invested in research and development, but also often implies that these fees are transferred to the licensees and to genetic products and technology users. This temporary monopoly over the cost of products can generate access barriers for those who are unable to pay. Needs which are not voiced by purchasing power on the market are not taken into consideration by the patent system. Therefore, the capacity to pay (or not) for a patented good can contribute to generate health gaps within and between countries.[197]

3.3.2.2 Patents and distributive justice in health for equality of opportunities

As we saw with the actual system, dissemination is often constrained by the willingness and capacity to pay for access and improved commercialisation of invention does not guarantee equitable access. This limitation on diffusion, access and utilisation of ideas can impact individuals' self-realisation and the progress of scientific innovation. Moreover, another problem with the way the actual system works is that the public is seen as a vague entity rather than a group of individuals who may have legitimate claims on patented knowledge and intellectual objects.[198]

---

195 Of course, some developing countries have manufacturing capacity and can produce goods domestically. However, with the universal standards established in TRIPS, those countries have to comply and award national protection to patent-holders, often to the detriment of their national generic industry. We will come back to the effect of patents on those countries in the last chapter.

196 Again, patent-holders' actions are obviously constrained by market forces, competition, and trade legislation.

197 This ability will depend on various factors. For example, some countries, like Canada, provide universal health insurance for certain products and services; many countries do not. Depending on whether the patented product is covered by an insurance programme or not, it can be each individual's business to assume their personal health care costs. D.G. Richards, *supra* Introduction, note 31, chapter 2, at pp. 25–38.

198 C. May, *supra* note 1.

Our global distributive justice theory demands that we do not harm global health; that institutional protection, relief, and aid development be provided against harm; and that genetic resources be redistributed to this end. These values and duties are not especially taken into consideration in the IP system which allows retention of property rights and control over inventions that, if otherwise accessible, could be crucial in meeting individuals' basic health needs. As Palmer notes, 'intellectual property rights, however, do not arise from scarcity, but are its cause'.[199]

In fact, property owners have the right to control access and decide how the object of their right will be used, produced, and exploited.[200] This necessarily influences the distribution of intellectual goods, but this alone does not say much about the quality of the distribution or its consequences for justice. We thus need to go one step further in enquiring about the compatibility of IPRs with global distributive justice. We already made our point on why we reject distribution undertaken by the free market or any *efficient* distribution that maximises the total amount of knowledge when it does not care about how this knowledge is distributed among individuals.[201] We instead argue for a mechanism that assesses distribution from a social welfare angle and grants direct help to individuals in need of health-related resources, technology, and services in order to bring them to a level where they can benefit from equality of opportunities. To this end, isolated charitable actions like temporary suspension of one company's drug patent in a particular country or drug donations in a few countries, are not sufficient to address the most vulnerable health needs. New innovative models and partnerships between industries, NGOs, governments and private foundations are a step in the right direction but should not take our attention away from a broad critique of the weaknesses of the global IP system when it come to distributive justice. In fact, obligations of distributive justice demand rethinking IPRs and global institutions from a theoretical point of view, and require constant action and commitment from the agents in charge of establishing and enforcing IPRs both at the national and the global levels. Distributive justice also demands rejecting the protectionist scheme of IP and avoiding artificial shortage in intellectual objects that does not allow prioritising the neediest.[202] We have seen that Rawls' view on distributive

---

[199] T.G. Palmer, 'Are Patents and Copyrights Morally Justified? The Philosophy of Property Rights and Ideal Objects' (1990) 13 *Harvard Journal of Law & Public Policy* 860, at 861.

[200] H.T. Engelhardt Jr., *The Foundations of Bioethics* (New York: Oxford University Press, 1986) at pp. 342–343.

[201] These views represent the libertarian and utilitarian perspectives on knowledge distribution, two philosophies discussed and critiqued in the theory chapter.

[202] See P. Drahos, *supra* chapter 1, note 56, at pp. 194–198.

justice suggests that inequality in distribution can be justified if it advantages the least well-off. I critiqued this justification because it still allows major disparities in access to genetic services and technology and is not concerned with equality, but with absolute improvement of the situation of the least well-off, even if it can be consistent with an important health divide.[203] Therefore, I instead focused on justifying distribution in terms of its effect on existing health needs and equality of opportunities.

Equality of opportunities in genetics implies that everyone should have access to a certain level of health in order to be in a position to take advantage of different opportunities available in crucial spheres of life, and therefore be capable of achieving their full potential. As discussed in the first part of this book, health is something of very special importance and universal significance for individuals of the world. I argued that, because of this, access to health requires particular standards of egalitarianism. In other words, as health is a vital element for every individual's personal development, the actual level of inequality in this area contributes to the preservation of a degrading and unfair level of inequality. This led me to conclude that inequalities in health should be repaired without taking individuals' personal financial situations into consideration. Hence, genetic knowledge, products, and services should be available and affordable to individuals in order to allow them to benefit from the different opportunities available. As more and more knowledge is produced, the amount of what is needed to be comfortable and capable of seizing opportunities increases as well. As discussed in the first part, even if the nature of opportunities is likely to vary between countries, the types of opportunities, such as the opportunity to pursue life and career undertakings, should be the same for everyone regardless of their nation, state, or ethnic group.[204] Taking this into account, property rights should be used as tools to maximise access to health and genetic-related knowledge, products, and services to ensure true equality in opportunities.[205]

However, this is not necessarily how the IPR system operates. In fact, patents can be awarded on genes, DNA sequences, tools, sequencing techniques, and many other important resources for moving genetic science ahead. Nevertheless, multiple overlapping patents can oblige researchers to pay high prices for several licences just to be allowed to conduct research projects, without any guarantee that they will result in positive outcomes. This is called

---

[203] D. Fallis, *Social Epistemology and the Digital Divide*, presented at the Computing and Philosophy Conference, Canberra, 2004, online on the website of the CRPIT: http://crpit.com/confpapers/CRPITV37Fallis.pdf (accessed 20 January 2009); H. Shue, *Basic Rights supra* chapter 2, note 63, at p. 128.

[204] H. Shue, *Basic Rights*, *supra* chapter 2, note 63, at pp. 59–60 and 159 et seq.

[205] See P. Drahos, *supra* chapter 1, note 56, at 178.

'royalty stacking' and it can influence availability in discouraging the use of patented genetic knowledge in further research and innovation. As discussed earlier, this is particularly true in areas of specific interest for people living in less profitable markets because of excessive research costs and small possibility of financial returns. Furthermore, patent-holders are free to charge licensing fees and high prices to licensees and users of genetics technology and products, mostly because of their need to earn back the capital invested in developing those products, and also to engage in further innovation endeavours. This means that people's capacity to afford *available* patented products and services determines access to genetic advances. Thus, IPRs awarded in genetics are especially important because they can have direct influence on individual health by creating or worsening differences in people's health within and among nations. Therefore, in influencing individuals' health status and constraints, IPRs play an important role in shaping the opportunity package from which individuals can choose.

Because health is an important prerequisite for taking advantage of available opportunities, property rights in genetics can play a large role in shaping what opportunities will actually be available to individuals. The intellectual property system leaves much latitude to private IP owners who can control how and by whom patented products and services will be used and, consequently, whose health level will improve the most. In the genetic sector, strong enforcement of IPRs can delay the availability of crucial health-related products and services, which can be evaluated though the death and diseases of the less affluent.[206] Hence, IP owners have some power over health status, and this influence can be positive or negative, depending on the philosophy and the licensing strategy they choose to adopt. In other words, the actual system gives IP owners the freedom either to help individuals around the world meet their health needs (by contributing to developing affordable innovative technology) or making it more difficult (or impossible) for the less affluent to access the technology they develop. With this perspective in mind, it can be hard to justify IP when some people are made *worse off* by not getting the same chances to improve their health. Indeed, when they cannot access patented products, people are not made worse off in the absolute sense, as nothing really changes: they did not have access to those products before they were invented and they still do not get access to them after. They are, however, worse off in a relative sense, since something that could improve their condition exists and they cannot access it because of strong patent rights. They are thus worse than they could actually be.[207]

---

206 C. May, *supra* note 31, chapter 4.

207 For a more detailed discussion on this point, see R.L. Ostergard, Jr., *supra* note 44, at 169.

Even if the IPR scheme generally tends to treat all intellectual goods the same without taking their role and importance in meeting basic needs into account, there are some broad flexibilities and health equity safeguards built into the international IP system, both in TRIPS and in the Doha Declaration.[208] A few examples: art. 7 of TRIPS mentions that IP protection should promote a balance between technological innovation and transfer and dissemination of technology for social welfare. Article 8 stipulates that states can adopt measures necessary to protect public health and nutrition, and to promote the public interest in sectors of vital importance to their socio-economic and technological development, but only if they are not contrary to the stipulations of the rest of TRIPS. Member states also have the freedom to exclude diagnostic, therapeutic and surgical methods for the treatment of humans or animals from patentability (art. 27(3)a) of TRIPS) and to award compulsory licences in limited cases (art. 31 of TRIPS).[209] Moreover, the Doha Declaration is meant to enable developing countries to pursue certain public health objectives, and states that countries can interpret TRIPS so it does not work against their health policies. It also allows states to grant compulsory licenses, especially to promote universal access to medicines, if they have sufficient manufacturing capacity in the pharmaceutical sector (art. 4); for those who do not, they can look for it in other countries.[210] Furthermore, art. 5d) of the Doha Declaration, in combination with art. 6 of TRIPS, allows member states to decide how they wish to enforce the principle of exhaustion of rights within their territory.[211]

---

208 J.H. Reichman, 'From Free-Riders to Fair Followers: Global Competition under the TRIPs Agreement', (1997) 29 *N. Y.U.L.J. Int'l. & Pol.* 11.

209 Compulsory licences enable a government to license a company, government agency, or other party the right to use a patent without the title holder's consent under strict conditions.

210 This has been made possible with the WTO 30 August 2003 decision to lift TRIPS restrictions on compulsory licensing and allow exportation of generic medicines to countries that are not in a position to manufacture them themselves. *WTO General Council, Decision on the Implementation of para. 6 of the Doha Declaration on TRIPS Agreement and Public Health*, 30 August 2003, online on the WTO website: http://www.wto.org/english/tratop_e/trips_e/implem_para6_e.htm (accessed 20 January 2009).

211 This means that countries can choose between national, regional, and international exhaustion of rights. For an interesting paper on the issues arising with parallel importing see: K.E. Maskus, *Parallel Imports in Pharmaceuticals: Implications for Competition and Prices in Developing Countries, Final Report to World Intellectual Property Organization*, Geneva, 2001, online on the website of WIPO: http://www.wipo.int/about-ip/en/studies/pdf/ssa_maskus_pi.pdf (accessed 2 February 2009).

Applying these types of measures could reduce the market price of health-related products and have a general positive effect on research and availability of genetic innovation. In reality, however, these dispositions are given the absolute minimum consideration by the community of people who interprets them. Developed countries and multinational corporations are putting strong pressure on developing countries wanting to use TRIPS' flexibility and, above all, are negotiating bilateral and regional free-trade agreements (FTAs) to impose more severe and contingent IP standards than those outlined in TRIPS.[212] Strong resistance from the most affluent countries toward any initiative from the developing world to take advantage of TRIPS and Doha's social welfare flexibility highlights the growing importance of strong and narrowly interpreted property rights on the global scene due to normative and political reasons.

I will first address the normative explanation for the poor application of Doha's and TRIPS' *flexibility clauses*. Although principles of the Doha Declaration have been adopted by the ministerial conference, the top decision-making body of the WTO, they remain very general and serve more as ethical guidelines (in the sense that they cannot be enforced in front of the WTO dispute settlement body). In addition, TRIPS' *flexibility clauses* contain important restrictions.[213] Article 30, for example, stipulates that every exception to the patentees' exclusive rights should be limited to ensure that they do not 'unreasonably conflict with a normal exploitation of the patent and do not unreasonably prejudice the legitimate interests of the patent owner'. Moreover, art. 31 establishes a long list of conditions that have to be met before compulsory licensing can be allowed in circumstances other than emergency and public non-commercial use, leaving many issues unsettled and much space for interpretation. As for the WTO's 30 August 2003 decision, it theoretically encourages cooperation between nations as a priority over protection of patent rights in certain extreme cases. However, it is associated with several conditions and many are expressing doubts about its practical applicability.[214]

---

212 Those agreements are referred to as TRIPS plus and are explicitly condemned by the WHO Commission on Intellectual Property Rights, Innovation and Public Health (CIPIH) in its April 2006 report: WHO Commission on Intellectual Property Rights, Innovation and Public Health, *Public Health, Innovation and Intellectual Property*, Geneva, April 2006, online on the WHO website: http://www.who.int/intellectualproperty/documents/thereport/CIPIHReport23032006.pdf (accessed 21 May 2006), rec. 4.21 and 4.25.

213 N.A. Bass, 'Implications of the TRIPS Agreement for Developing Countries: Pharmaceutical Patent Laws in Brazil and South Africa in the 21st Century' (2002) 34:1 *George Washington International Law Review* 191.

214 For example, see J. Lanjouw, *Complementarity of my FFL proposal and Canada's approach in its Pledge Legislation (C- 9): Comments for the CIPP Forum*, Montréal, January 2005.

This leads us to the political explanation of why the flexibility embodied in TRIPS and Doha is not translated in concrete results. In fact, because of the numerous conditions that countries in need have to meet to benefit from TRIPS' exceptions, and because of the existing space for competing interpretations, bargaining power has become crucial in establishing the scope of these exceptions. Needless to say, this exercise almost always benefits the most affluent countries and powerful stakeholders, who tend to object strongly to the use of these exceptions, often threatening to impose trade sanctions against countries that express interest in engaging in them.[215] More importantly, there is a strong tendency from developed countries to impose even stricter standards on developing countries with bilateral and regional free-trade agreements. Moreover, powerful countries and corporations not only try to prevent developing countries from using available flexibility in practice, but they also closely monitor how they construct their domestic patent laws.[216]

All of this clearly illustrates Drahos' view on proprietarianism and the fact that it occupies a large place in how the actual IPR system works and is justified. As it currently stands, the world economy is based on strong property rights mainly driven by market forces. This results in huge power concentration in the hands of a few elites who can run the system and prioritise their own values and interests. As discussed earlier, even if it should theoretically further a balance between protection and diffusion other than imposing mandatory public disclosure, the IP system does not impose enforceable and

---

[215] In fact, no generic drugs have been produced using compulsory licenses to treat patients in the last decade. Brazil is the only example of a country that resisted pressure and successfully used the threat of compulsory licensing in its price negotiations with pharmaceutical companies in the context of its national AIDS strategy. Brazil was in a position to do so because of its important research and manufacturing capabilities. However, very few countries are in this negotiating position, as appears from an investigation conducted in 2001 in about 70 developing countries, which found that only half of them were providing for international exhaustion of patent rights in their domestic patent legislation. UK Commission on Intellectual Property Rights, *supra* Introduction, note 40, chapter 8, at 160. See also M.M. Nerozzi, 'The Battle Over Life-Saving Pharmaceuticals: Are Developing Countries Being "TRIPped" by Developed Countries' (2002) 47 *Vill. L. Rev.* 605.

[216] This is illustrated by the famous 2001 law suit filed by a consortium of the biggest drug companies against the South African government. Those companies wanted to challenge the 1997 South African patent legislation (which allows the government to manufacture and import cheaper retroviral AIDS drugs), maintaining that it is too broad and unfair for brand name drug producers. Those major multinational pharmaceutical companies thus launched this law suit as part of a strategy to strongly encourage and pressure developing countries to adopt stricter patent protection standards. However, with worldwide public pressure and massive outrage raised by the consortium's action, the pharmas were left with no choice but to back off and drop the case.

demanding corollary obligations on IP owners in exchange for the rights awarded. The proponents of strong IP rights do not see any problem or contradiction in this, as they view the return on investment as a means to promote general welfare through further investment in other research endeavours. But, in reality, IPRs foster the interests of patentees and the more affluent in priority, without worrying too much about the inequities they create and encourage, or the consequences they can have on the lives and health of individuals.[217] As May summarises, '[t]he knowledge structure ensures that, as science is commercialised, property based mechanisms are introduced because they are common sense in market transactions'.[218]

For our goal of distributive justice for global health, we would need to replace this proprietarianist view with an instrumentalist attitude that would support a different social role for IP. Intellectual property rights should thus be conceived as tools to support better and broader access to health by every individual through principles of global distributive justice, instead of as a fixed system controlled by a few companies to support their own economic interests. Ownership should not be available when it works to exclude individuals from accessing crucial health-related goods and services, to delay their availability, and, consequently, to tacitly support more death and illness. Our cosmopolitan focus would instead require considering each individual, regardless of their country of origin, as a unit of moral concern for access to health – not leaving the result to an economic battle between their government representatives and the world's most powerful agents. Until the international IP scheme stops functioning primarily to further innovation and comes to care about ensuring that the results of such innovation are diffused so as to reach those who need them throughout the world, we will not be able to conclude that IP works to advance global distributive justice principles to further access to common health standards and allow real equality of opportunities.

Nevertheless, there are signs that things might slowly be changing in some business sectors despite the main focus, politics and governance of IP. Indeed, new business models are being explored by some industries and surprising partnerships are slowly emerging In this respect, it appears relevant to say a few words on international initiatives undertaken to address neglected diseases affecting the most-vulnerable and less-affluent of the world. One of them is UNITAID's medicine patent pool initiative which design implies a collaboration between brand name pharmaceutical companies, universities, generic companies, national governments and NGOs to facilitate access, availability, licensing and production of patented and non patented anti-retroviral drugs at

---

217 See P. Drahos, *supra* chapter 1, note 56, at 195–203.
218 C. May, *supra* note 31, chapter 4, at 117.

a reasonable cost for the needs of developing and emerging economies.[219] Another one is the partnership between the Drug for Neglected initiative (DNDi), a non-profit organisation involved in developing new drugs for neglected diseases and major brand name pharmaceutical compagnies, who have agreed to team up with the organisation. As a result of one successful partnership between DND and Sanofi, a new cheap combination drug to treat malaria was made available in March 2007 outside of the patent regime.[220]

However, despite those encouraging partnerships and notwithstanding what CEOs and managers of brand name pharmaceutical companies have recently said about changing their business models and not necessarily being into strong pharmaceutical patent enforcement anymore,[221] the lack of access to essential drugs in the developing world is still a huge issue.[222] Many non-legal, country-specific elements (such as physical infrastructure, education levels, and the political situation) can pose significant hurdles to accessibility in developing countries. Nevertheless, it is important to mention that there are a growing number of patents in areas connected to fundamental needs.[223] Although many essential medicines on the WHO list are not patented, most of the existing HIV/AIDS medicines are patented in numerous developed and developing countries and some very efficient drugs that can treat tuberculosis, and malaria have been patented and strongly enforced in recent years.[224]

---

[219] A patent pool arises when patent-holders mutually agree to license their patents to each other and to third parties.The UNITAID patent pool is not established yet but UNITAID was given the mandate to create it in July 2008. For more on this project, refer to UNITAID, *The Medicine Patent Pool Initiative, Fact Sheet*, (March 2009), http://www.unitaid.eu/images/projects/ppinfo.pdf (accessed 20 May 2009); see also E.R. Gold et al. *Preliminary Legal Review of Proposed Medicines Patent Pool* (26 July 2007), prepared by IPDS for UNITED.

[220] D.G. McNeil Jr., 'Low-Cost Antimalaria Pill Available' (1 March 2007), *The New York Times*, online http://www.nytimes.com/2007/03/01/health/01malaria.html?_r=2&ref=health&oref=slogin (accessed 20 May 2009).

[221] Groupe international d'experts en biotechnologie, innovation et propriété intellectuelle, *supra* note 70, at 20; Yves Mamou, 'Le lancement de nouveaux médicaments est de plus en plus coûteux et rapporte de moins en moins: Les laboratoires sont contraints de révolutionner leur recherche' (3 January 2008) *Le Monde* at 10.

[222] See Groupe international d'experts en biotechnologie, innovation et propriété intellectuelle, *supra* note 70.

[223] Commission on Human Rights, *Access to Medication in the Context of Pandemics such as HIV/AIDS, Resolution 2001/33*, in Report on the 57th Session, 19 March–27 April 2001, UN Doc. E/2001/ 23-E/CN.4/2001/i67.

[224] Patents do play a role in access to medicines: P.G. Harris, P. Siplon, 'International Obligation and Human Health: Evolving Policy Responses to HIV/AIDS' (2001), 15: 2 *Ethics and International Affairs*, at p. 29: H. Hestermeyer, *Human Rights and the WTO: The Case of Patents and Access to Medicines* (New York, Oxford University Press, 2007) at p. 208; M. Boylan, 'Medical Pharmaceuticals and

Prices are therefore affected by licensing fees that can have direct negative impacts on drug affordability and on the production of cheaper generic options.

It will therefore be very interesting to follow the new initiatives and see if they will contribute to build sustainable and efficient global health strategy or if, as some think, private companies will end up going back to their 'old' business models after using those partnerships as a public relations and marketing exercise.[225]

CONCLUSION

In this chapter, I analysed the international normative IP system to determine whether its underlying philosophy, structure, and functioning adequately account for the values encountered by our global distributive justice framework. My main goal was to assess the patent system to find out if it can facilitate the redistribution of potential genetic benefits, taking health needs into consideration.

All through the chapter, I tested the international IP system with my benchmarks of justice through different access lenses. I began with the global aspect of access to genetic resources and realised that the application of strong and broad patent rights, particularly in this field, was more compatible with the reduction of the public commons, with the creation of some health gaps associated with people's capacity to pay, and with an international basic structure established by a few stakeholders for their own benefit – and not to support principles of justice, shared global health ideals, or universal consideration of every human being.

I pursued my analysis in assessing the compatibility of IP rights with access to genetics, this time in relation to availability and affordability of genetic products and services. I again noticed that the current application of most patents was very strict and primarily market-driven, geared towards protection more than diffusion, and establishing artificial shortages in intellectual goods.

My overall analysis confirms Drahos' critique of the proprietarianist version of IP. In fact, I conclude that one of the biggest problems with the

---

Distributive Justice' (2008), 17, 30–44 *Cambridge Quarterly of Healthcare Ethics*, at p. 31.

[225] For more on this position, refer to B. Brubaker, 'The Limits of $100 Million; Epidemic's Complexities Curb Impact of Bristol-Myers's Initiative' (29 December 2000) *Wash. Post*, at A1; Oxfam, *Generic Competition, Prices and Access to Medicines: The Case of Antiretrovirals in Uganda,* Briefing Paper 26, 10 July 2002, online on the Oxfam website: http://www.oxfam.org.uk/what_we_do/issues/health/bp26_generic.htm (accessed 30 May 2006).

international IP system in relation to global access is that intellectual property rights are viewed as private ends in themselves rather than as tools to further the public interest in accessing new knowledge and encouraging innovation. The politics of IP is characterised by the powerful defenders of strict and protectionist IP standards on one side, and the less powerful on the other side who campaign for more focus on the public and social welfare aspects of the IP system.[226] The strongest and most powerful actors, both politically and economically, run the system now. They mostly adopt strict enforcement strategies against the poorer majority, going as far as imposing more and more demanding conditions in order to further their economic values and interests.[227] Even if we are witnessing some interesting changes in this regard, most major stakeholders have no interest in voluntarily getting involved in a market with poor potential for return, even if it could result in important human benefits. After all, their main reason to be in business is to make profit. However, it creates a vicious circle, as 'the failure to address the health care needs of poor people is to permanently consign them to both illness and poverty'.[228]

This is what makes the system, as it functions now, incompatible with my global distributive justice framework. If the IP system were to meet my goal of global distributive justice for global health, it would need to adopt a totally different social role, more focused on diffusion and needs than on production and protection of innovation, supporting broader access to health and considering every individual, not just the inventors, as units of moral concern. IPRs can play an important role through knowledge and innovation diffusion in allowing individuals to reach the health level they need to be able to profit from available opportunities. International trade and IP agreements will remain priorities for powerful industrialised countries but, as they currently operate, they are not working to ensure equitable access to health. Until there is a major change in IP philosophy and politics to allow social welfare concerns to be taken into consideration on a general scale (not just through a few high profile partnerships), we will not be able to conclude that IP works to advance global distributive justice principles towards access to common health standards.

As more and more people come to realise that taking care of the more pressing global health issues is critical for the creation of a more just and stable world order, coming up with balanced and fair IP mechanisms appears

---

226 S.K. Sell, *supra* note 34.

227 These agreements reproduce TRIPS-plus standards. To consult some of those agreements, refer to the website of the Office of the US Trade Representative, online: http://www.ustr.gov/ (accessed 19 February 2009).

228 D.G. Richards, *supra* Introduction, note 31, chapter 6, at p. 160.

to be one of the many important steps in the right direction. While TRIPS cannot and will not disappear, it is important for developing countries to be able to take advantage of its existing mechanisms to help them focus on their health priorities and, most importantly, to present strong and coordinated opposition to expansion initiatives. Thus, the growing post-TRIPS involvement of international NGOs, of private foundations and of a strong African leadership on health care issues will continue to be crucial in framing IP issues differently and tempering the industry-dominant influence over the IP agenda.[229]

As we saw at the beginning of this book the benefits arising from genetics have a real potential for improving global health; much may depend on how widely they get distributed. This does not mean that IP protection should be abolished altogether, as it can be an important trade mechanism for managing innovation and, if used and balanced adequately, it can play a major role in achieving better global health equity.[230] However, finding the right balance in the application of IP protection to genetics is an important challenge. Indeed, the system should, at the same time, prevent the more vulnerable from being left out of progress with diffusion while preserving a certain level of protection for inventors. Much more needs to be done to get to this point, either within the actual IP system or outside of it, and many different strategies have been proposed to this end.[231] I will come back to practical policy options at more length in the conclusion of the book to set some basis for further reflection.

The next chapter will focus on international human rights law. The purpose of this chapter will be to present another international normative system and to determine if its structure and functioning adequately account for the values encountered in our global distributive justice framework.

---

229 S.K. Sell, *supra* note 34; P. Drahos, *Developing Countries and International Intellectual Property Standard Setting*, UK Commission on Intellectual Property Rights, Study Paper 8, London, 2002, at p. 26.

230 E.R. Gold et al., *supra* chapter 1, note 54, at 312.

231 For an interesting and complete analysis of the role of IP in innovation and recommendations towards improving the role of IP and moving from Old IP to a New IP era, I refer the reader to the Groupe international d'experts en biotechnologie, innovation et propriété intellectuelle, *supra* note 70. See also L.R. Helfer, 'Regime Shifting: The Trips Agreement and New Dynamics of International Intellectual Property Lawmaking' (Winter 2004) 29 *Yale J. Int'l L.* 1, at 58.

4. International human rights law: a second tool?[1]

DO INTERNATIONAL HUMAN RIGHTS HELP OR HINDER THE REALISATION OF BENEFIT SHARING?

INTRODUCTION

The first part of the book set a theoretical framework to support equitable access to health, and more specifically distribution of genetic research benefits and resources to come. For this purpose, I argued for equal and universal consideration of every individual's basic health needs in support of a rationale to secure equal opportunities for all on the global scene. The second part of the book is dedicated to the assessment of two normative systems using the parameters established by my theoretical framework. In the last chapter, I assessed the intellectual property law system. The purpose of this chapter is to determine if the basis, functioning, and conceptualisation of the existing international human rights (IHR) legal system, especially socio-economic rights, helps or obstructs the realisation of global distributive justice in health.

I will start with a brief introduction to the system of IHR and its main philosophical foundations. The second part of this chapter will assess IHR law with reference to notions of equality, global distribution, and justice, once again using the standard of access. Finally, in the third and last part of this chapter, I will provide an analysis of how human rights are conceptualised through the reality of the market.

---

[1] Sections of this chapter were inspired by previous work published in the *International Journal of Human Rights* (Louise Bernier, 'International Socio Economic Human Rights: The Key to Global Health Improvement?' 14:2 *International Journal of Human Rights* (forthcoming, Spring 2010).

4.1 THE FIELD OF IHR LAW

4.1.1 Presentation

The notion of human rights emerged from the need for universal respect for human beings' freedom, dignity, and equality. It has been translated into a common language and set of identified human rights. This system is the result of a long struggle to gain universal support for individual protection from oppression, and to give all people an equal chance to develop their potential to be able to take advantage of different opportunities.[2] Indeed, since the Enlightenment, various human rights claims resulted in slowly liberating individuals and communities from repressive regimes and institutions,[3] but the human misery and atrocities that have happened during and following the Second World War are viewed as having been a turning point in the reintroduction of the modern ideal of human rights.[4] Indeed, the *Universal Declaration of Human Rights* (UDHR) adopted by the UN General Assembly in 1948 marked a crucial step in the effort of the international community to establish a common standard of achievement for all peoples and all nations.[5] Together with the UDHR, the *International Covenant on Civil and Political Rights* (ICCPR) and the *International Covenant on Economic, Social and Cultural Rights* (ICESCR) form what has been called the *International Bill of*

---

[2] J. Donnelly, *Universal Human Rights in Theory and Practice* (London: Cornell University Press, 1989).

[3] Indeed, some of the most important efforts in the development of human rights in the late-seventeenth and eighteenth century are the *British Bill of Rights* (1690–91), the *American Declaration of Independence* (1776), and the *French Declaration on the Rights of Man and the Citizen* (1779). A. Eide, 'Economic, Social and Cultural Rights as Human Rights' in A. Eide et al. (eds), *Economic, Social and Cultural Rights: A Texbook*, 2nd edn, (London: Maartinus Nijhoff, 2001) p. 9.

[4] T. Ball and R. Dagger, *Ideals and Ideologies: a Reader* (New York: Pearson Longman, 2004).

[5] The early 1960s was marked by a new wave of human rights activity, which led to various international documents, including, for example, the 1959 *Declaration of the Rights of the Child*, the *European Convention on Human Rights* of 1963, the *International Convention on the Elimination of All Forms of Racial Discrimination* of 1965, the *International Covenant on Economic, Social, and Cultural Rights*, and the *International Covenant on Civil and Political Rights* (opened for signature and ratification in 1966 and coming into force in 1976), the 1975 *Declaration on the Rights of Disabled Persons*, and the 1979 *Convention on the Elimination of all Forms of Discrimination Against Women*. For international legislation on human rights, see I. Brownlie, *Basic Documents on Human Rights*, 3rd edn (Oxford: Clarendon Press, 1992).

Human Rights.[6] The conception of modern human rights can be described as rights recognised for all individuals simply because they are human beings. Human rights aim to enhance people's quality of life and involve a direct link between them and their state, whose actions should be in conformity with the established international human rights regime.[7] Indeed, even if human rights are primarily established on the international scene, they also have to be developed at the national level, where they become legally enforceable.

There are two main categories of human rights. First, civil and political rights, including democracy, due process, and freedom of expression, which have to be guaranteed immediately by signatory states. Secondly, economic, social, and cultural rights, which can be realised progressively, and which can include rights to social security, work, adequate food, and the highest attainable standard of health. Although they have not yet been accepted as legally enforceable, there is increasing support for an emerging third category of human rights called solidarity rights, asking for more equitable distribution and protection of common resources to benefit not only individuals, but also communities at the international level.[8]

As mentioned earlier, the notion of rights is often criticised on the basis that rights can remain meaningless in cases where taking advantage of them is not a real option for the right-holders. This objection clearly highlights the need for a better interconnectedness between socio-economic rights and civil and political rights. Indeed, even if it seems quite easy to grant rights like the right to vote, the right to freedom and security, or the right to free speech, if personal health, material and economic conditions of the right-holders are such that it is literally impossible for them to profit from those rights, they become mean-

---

6 Contrary to the UDHR, the two Covenants are legally binding on the states that ratified them, who thus have to comply with the treaties' provisions and submit reports on concrete actions taken toward this end. *International Covenant on Civil and Political Rights* (ICCPR) G.A. res. 2200A (XXI), 21 U.N. GAOR Supp. (No. 16) at 52, U.N. Doc. A/6316 (1966); *International Covenant on Economic, Social and Cultural Rights* (ICESCR), G.A. Res. 2200 (XXI/, UN GAOR, 21st Sess., Supp. No. 16, at 49, UN Doc. A/6316 (1966).

7 E.A. Andersen and B. Lindsnaes, *Towards New Global Strategies: Public Goods and Human Rights*, Dordrecht: Martinus Nijhoff Publishers, 2007) at p. 66; J. Mann et al., *Health and Human Rights* (New York: Routledge, 1999) chapter 1, at p. 8; art. 1 of the Vienna Declaration and Programme of Action, UN Doc. A/CONF.157/24 (part 1).

8 This third category refers, for example, to a right to peace, to a clean environment, to benefit from the common heritage of humankind, and to development. For more on solidarity rights, refer to P. Alston, 'A Third Generation of Solidarity Rights: Progressive Development or Obfuscation of International HR Law' (1982) 29:3 *Netherlands International Law Review* 307.

ingless. As Pogge clearly states, 'in a situation where there is formal freedom but extreme poverty, the poor are in many obvious ways unfree on account of their poverty'.[9] Even if economic, social, and cultural rights (the human rights most frequently breached) are highly criticised for their lack of justiciability and conceptual precision,[10] it might make more sense to eliminate what is now a clear separation between civil and political rights and economic, social, and cultural rights.[11] This is consistent with guideline 4 of the 1997 *Maastricht Guidelines on Violations of Economic, Social and Cultural Rights*, which states that '[i]t is now undisputed that all human rights are indivisible, interdependent, interrelated and of equal importance for human dignity. Therefore, states are as responsible for violations of economic, social and cultural rights as they are for violations of civil and political rights.'[12] Consequently, since poverty and ill health are important obstacles to a satisfying human existence as established earlier, economic and social rights should rank alongside civil and political rights,[13] and all of these rights should thus be envisioned as essential values for a better world.[14] However, although I acknowledge the indivisibility, interdependence and interrelation among rights, it is important to note that my main focus will be on the content, conceptualisation, and realisation of socio-economic rights as I argue for global distribution and access to genetics research benefits in support of better health. Now, before getting to the core analytical part of the chapter, I will briefly present the two main discourses on which human rights are based.

---

9 T. Pogge, *Realizing Rawls* (Ithaca: Cornell University Press, 1989), at 133.

10 See for example B. Toebes, 'Towards an Improved Understanding of the International Right to Health' (1999) 21:3 *Human Rights Quarterly* 662; B. Tolchin, 'Human rights and the requirement for international medical aid' (2008) 8:2 *Developing World Bioethics*, at 151.

11 The strong connection between the two categories of rights and the fact that there is no hierarchy between them has been specifically emphasised by the UN: United Nations Committee on Economic, Social and Cultural and Economic Rights, *General Comment No. 3(1)*, 2002, online on the website of the UN: http://www.unhchr.ch/html/menu2/6/cescr.htm (accessed 1 April 2009).

12 Maastricht Guidelines on Violations of Economic, Social and Cultural Rights, Guideline 4 (1998) 20 *Human Rights Quarterly* 692.

13 B. Terence and R. Dagger, *Political Ideologies and the Democratic Ideal*, 5th edn (New York: Addison Wesley, 2004).

14 J. Waldron, *supra* chapter 2, note 71, at pp. 7–8.

4.1.2 Theoretical and Legal Foundations of IHR

The existence and validity of human rights are not written in the stars... [They] have been conceived and taught by enlightened individuals in the course of history.[15]
Albert Einstein

The nature and underlying justification of the concept of rights have undergone considerable change over time.[16] As mentioned previously, the view of rights put forward by Hegel, Kant, and other philosophers during the nineteenth and the twentieth centuries was heavily criticised for reducing the concept of rights to *social constructs*.[17] This explains why rights were absent from the vocabulary of political philosophy for decades before coming back, after the Second World War, in legal and philosophical discourse.[18] This section seeks to present briefly those two main discourses and understand how they sometimes complement and contradict each other while also giving rise to debate outside their respective frameworks.

4.1.2.1 The legal discourse of IHR

As mentioned in the previous section, an important body of IHR instruments has been established on the global scene and further developed at the national level. This highlights an inherent conflict between, on the one hand, a universal and cosmopolitan way of envisioning human rights and, on the other hand, the notion of state sovereignty, a basic principle of international law. This can be resolved when states agree to concede some portion of their sovereignty by ratifying international norms. Human rights are often envisioned as a basis for justifying demands and imposing obligations. Indeed, one purpose of this legal system is to ensure that a right-holder can demand that the content of his or her right be guaranteed. Rights are meant to allow a connection between human well-being and related obligations. Because of the way the system is built, individuals' rights are directly related to the status of citizen. In other words, individuals can take advantage of human rights only when their states have decided to recognise and enforce those rights based on a sense of 'justified

---

[15] A.P. French, *Einstein: a Centenary Volume* (Cambridge: Harvard University Press, 1979) p. 305.

[16] For a comprehensive analysis of the evolution of the concept of right, see G.B. Herbert, *supra* chapter 2, note 81.

[17] J. Bentham, *supra* chaper 3, note 54; K. Marx, 'Capital', in F. Engels (ed), *Manisfesto of the Communist Party,* Translated from the 3rd German edn by S. Moore and E. Aveling. Rev (Chicago: Encyclopædia Britannica, 1955); G.B. Herbert, *supra* chapter 2, note 81, at p. 277.

[18] G.B. Herbert, *supra* chapter 2, note 81, at p. 286.

outrage and political empowerment'.[19] These rights are included in legal instruments that serve to recognise the fundamental and general political values agreed to by the international community. They place individuals at the centre of national and international legal concerns, with a clear recognition of the concepts of human respect and dignity.[20] The purpose of the legal discourse is not to question the content, the essence, or the interests that lie behind these norms, but instead to recognise, clarify, and enforce the general legal rules emerging as a product of international law. As Evans notes, '[t]he legal discourse focuses upon the internal logic of the law, its elegance, coherence, extent, and meaning, which the application of legal reason is said to reveal'.[21]

Many scholars are opposed to the legal discourse of human rights for a variety of reasons. First, many talk about the numerous deficiencies in the legal mechanisms needed to support such legal discourse. Indeed, solid legal infrastructures, strong enforcement strategies, and real sanctions are still lacking in IHR law.[22] This position is clearly explained by Kennedy when he says: '[t]he attachment to rights as a measure of the authenticity, universality, and above all as the knowledge we have of social justice binds our professional feet, and places social justice issues under the governance of the least effective institutional forms available'.[23]

Others are opposed to this discourse because it seems to conceal an arrogant assumption of what is good for people, encouraging an empty sense of entitlement on the part of individuals, all of which has the perverse effect of fostering passivity among individuals and communities. In other words, some

---

19 K. Robinson, 'False Hope or a Realizable Right? The Implementation of the Right to Shelter under the African National Congress' Proposed Bill of Rights for South Africa' (1993) 28 *Harvard Civil Rights-Civil Liberties Law Review* 505, at 517.

20 W. Austin, 'Using the Human Rights Paradigm in Health Ethics: The Problems and the Possibilities' (2001) 8:3 *Nursing Ethics* 183; Report of the United Nations High Commissioner for Human Rights, A/53/372, 11 September 1998, New York.

21 T. Evans, 'International Human Rights Law as Power/Knowledge' (2005) 27:3 *Human Rights Quarterly* 1046, at 1050.

22 In fact, even if some international UN agencies are responsible for promoting human rights, there is no consistent monitoring, reporting, or enforcement practice, and only very weak judicial and quasi-judicial activities in this field. This explains the vagueness and deficient conceptual clarity of many of these legal rights. For more information on this topic, refer to A.R. Chapman, 'Monitoring Women's Right to Health Under the International Covenant on Economic Social and Cultural Rights' (1995) 44 *Am. U. L. Rev.* 1157, at 1159–1160; Kate O'Regan's judgment in *S v Makwarryane* 1995 (3) SA 391 (CC), 1995 (6) BCLR 665 (CC) at para 325.

23 D. Kennedy, 'The International Human Rights Movement: Part of the Problem?' (2002) 15 *Harvard Human Rights Journal* 101, at 140.

critique the very institution of granting legal human rights because it can (wrongly) incite a sense of accomplishment in governments and stasis in individuals who theoretically have individual legal rights, but can hardly enforce them. For opponents of legal human rights, this system seems to encourage focus on complex and deficient legal dilemmas and institutional procedures which gives the illusion of control over human rights violations instead of concentrating actions and efforts on better and more inclusive modes of action.[24]

A third category of opposition to the legal discourse concerns the limits of legal positivism and the fact that it precludes any deeper analysis of what lies at the source of the norms encoded in the law. In fact, it can be argued that the system endorses a kind of realism regarding established powers – what has been called engagement with the *realpolitik* of human rights.[25] This can limit the means to address abuses in preventing broader political, economic, and philosophical analysis of right violations.[26]

These criticisms highlight certain limitations of the dominant legal discourse of human rights and illustrate the importance of serious inquiry about the real nature and justification for the principles enshrined in those legal instruments. This brings us to a different, less legal and institutional way of envisioning human rights within a theoretical context.

4.1.2.2 The moral discourse

As highlighted by various scholars, there is a tension between the legal and moral discourses on rights.[27] The moral discourse on human rights refers to a justification based on more profound and objective reasons than what the legal approach offers. Theoretically, we can say that human rights are recognised and understood as having been founded on the basic interests that individuals have in the content of those rights. As explained earlier, an argument for a right 'is an argument showing that an individual interest considered in itself is sufficiently important from a moral point of view to justify holding people to be

---

[24] Ibid.

[25] N. Stammers, 'Social Movements and the Social Construction of Human Rights' (1999) 21:4 *Human Rights Quarterly* 980, at 991.

[26] T. Evans and E. Hancock, 'Doing Something Without Doing Anything: International Human Rights Law and the Challenge of Globalisation' (1998) 2:3 *Int J Hum Rights* 1.

[27] C. Brown, 'Universal Human Rights: A Critique' in T. Dunne and N.J. Wheeler (eds), *Human Rights in Global Politics* (Cambridge: Cambridge University Press, 1999) chapter 3, p. 103; P. Jones, *Rights* (London: Macmillan, 1994); Michael Freemen, *Rights* (Minneapolis: University of Minnesota Press, 1991). See also A. Buchanan, 'Human Rights and the legitimacy of the international order' (2008), 14 *Legal Theory*, 39.

under a duty to promote it'.[28] Some criticise the human rights system because it seems to lack a uniform ideal of what should count as a human right.[29] Many different theories can act as foundations for human rights claims, the most important being natural law, which focuses on neutral values like individual freedom, equality, and universalism.[30]

Brown states that any 'idea of natural law must underlie all genuinely universal approaches to human rights'.[31] Natural law implies that the emerging principles are not related to specific types of societies, institutions, or enforcement procedures but arise instead from the protection of characteristics associated with human agency or personhood. Features of agency have been regrouped in four categories: (a) capacity to make life decisions without undue pressure; (b) ability to acquire a certain basic level of education and knowledge; (c) capacity to undertake some chosen projects; and (d) liberty to pursue what one perceives as the good life.[32] This supports our cosmopolitan view that all individuals are equal members of a single moral universe, that they all have dignity, and that all require fulfilment of similar basic conditions to be able to grow and live a good and dignified life. Therefore, all individuals have human rights because they are equal human beings, and rights can serve to ensure that they will to be able to create what they wish and pursue certain life goals. It does not mean that any injustice constitutes a violation of human rights, but only that when it touches elements of human agency, such as health, special protection is needed to ensure that individuals benefit from equal treatment and have access to basic requirements to be able to pursue personal life goals and take advantage of available opportunities.[33]

One important step in establishing a theoretical discourse of human rights is thus to identify which characteristics of human nature deserve special protection and which values should be respected everywhere. Some argue that rights should be ranked according to the nature of the interest they help defend

---

[28] J. Waldron, *The Right to Private Property*, *supra* chapter 2, note 71 at p. 3. On the interest notion see also: J. Waldron, *supra* chapter 2, note 71 at pp. 11 and 212; J. Raz, *The Morality of Freedom* (Oxford: Clarendon Press, 1986) at p. 166.

[29] For example, see J. Griffin, 'Discrepancies Between the Best Philosophical Account of Human Rights and the International Law of Human Rights' (2001) 101 *Proceedings of the Aristotelian Society* 1.

[30] For more on the concept of natural rights, see C. Brown 'Universal Human Rights', *supra* note 27; T. Evans, 'Universal Human Rights: As Much Round and Round As Ever Onward' (Winter 2003) 7:4 *The International Journal of Human Rights* 155; J. Finnis, *Natural Law and Natural Rights* (Oxford: Clarendon Press, 1980), at p. 23.

[31] C. Brown, 'Universal Human Rights', *supra* note 27, at p. 106.

[32] J. Griffin, 'First Steps in an Account of Human Rights' (2001) 9:3 *European Journal of Philosophy* 306.

[33] Ibid.

and their normative weight in the process of reaching a stage where individuals can function as human agents taking advantage of equal opportunities.[34] Others believe in the protection of basic rights associated with the primary necessities and preservation of human life. Such basic rights can emerge from the basic needs shared by every human, such as food, water, shelter, and health care.[35]

Some relativists criticise the underlying universality of the international system of human rights; they argue that there is no true universality or universal community, and that human nature and basic moral principles are instead constructed by external factors like history and culture. They argue that this would explain the existing important variation between moral practices around the world.[36] Others criticise the individualistic aspect of human rights on the basis that it seems to encourage the construction of individuals outside any form of community and to support a disturbing and harmful individualism typically observed in the western world.[37] We will come back to those criticisms in the next section when we assess the universalism of human rights principles.

An important point of a moral account of human rights relates to the nature and content of correlative duties. As we mentioned previously in our theoretical chapter, some, like O'Neill, go as far as making the very existence of a right depend on correlative duties to respect and fulfil those rights. They consequently express considerable scepticism toward welfare human rights due to implementation challenges. In other words, for those opponents, although rights-talk is rhetorically powerful, it is not ethically founded because it does not deal with the powerful actors who could do something about international injustice. Others, like Shue and Pogge, differentiate between positive duties and negative duties Shue ranks human rights protection and puts basic rights at the top of his list. He establishes special responsibilities to fulfill the most urgent and serious rights without necessarily taking responsibility for rights violation into account.[38] Pogge, on the other hand, argues that obligations should be linked to responsibility for deprivation, giving rise to a duty to avoid causing harm to others. He associates human

---

34 O. O'Neill, 'Hunger, Needs and Rights', *supra* chapter 1, note 22; P. Jones, *Rights* (London: MacMillan, 1994), at pp. 13–15.

35 H. Shue, *Basic Rights, supra* chapter 2, note 63.

36 K. Booth, 'Three Tyrannies' in T. Dunne and N. Wheeler (eds), *Human Rights in Global Politics* (New York: Cambridge University Press, 1999) p. 31.

37 For more on this topic, we encourage the reader to refer to these classics: J. Waldron, *Nonsense upon Stilts, supra* chapter 2, note 82; K.H. Marx, *On the Jewish Question* (written in 1843); G.W. Hegel, *Hegel's Philosophy of Right* (T.M. Knox, trans.) (Oxford: Clarendon Press, 1964).

38 H. Shue, *Basic Rights, supra* chapter 2, note 63, at pp. 52 and 118.

rights protection with a negative duty not to uphold an unjust international order, and believes that everyone who supports the unjust global structure is responsible for the human rights violations it causes.[39] We will not say more on the question of correlative duties for now as they will be analysed at length in sub-section 4.2.2.2 dedicated to human rights implementation and access to health.

A different way of envisioning a philosophical account of human rights, instead of questioning the moral foundation of the very notion of human rights, is to refer to the concept as part of a theory of justice. As Kymlicka clearly puts it, 'justice can be considered as the system of entitlements upon which people can base their demand for recognition of their legitimate claims for resources and opportunities'.[40] This allows human rights to exist without actual claims of specific duties against identifiable actors, as they are associated with the ideal justice values and principles to which we aspire globally.[41] Rawls, for example, envisions human rights as a component of his theory of justice, as an important aspect of what should drive the interaction between different states, and also between citizens and their government. In fact, respect for human rights is one of the eight principles of the law of peoples, principles that should be followed by every decent society. For Rawls, human rights establish the limits of what should be tolerated from other societies, and are common to every decent society, liberal or not.[42] This narrow definition justifies including only a few rights in the list – such as rights to life, liberty, freedom of expression and religion, property, and equality before the law – and excluding most equality and welfare rights.[43]

---

[39] T.W. Pogge, 'Symposium on World Poverty and Human Rights', *supra* chapter 2, note 97; T. Pogge, 'Eradicating Systematic Poverty: Brief for a Global Resources Dividend' (2001) 2 *Journal of Human Development* 59; T. W. Pogge, 'Reponsibilities for Poverty-Related Ill Health' (2002) 16:2 *Ethics and International Affairs* 71, at 73; L. Doyal and I. Gough, *supra* chapter 1, note 42.

[40] W. Kymlicka, *Liberalism, Community and Culture*, *supra* chapter 2, note 69, at 234.

[41] A. Gewirth, *supra* chapter 2, note 92, at 330.

[42] C.R. Beitz, 'Human Rights as a Common Concern' (2001) 95:2 *American Political Science Review* 269; A. Buchanan, 'Rawls's Law of Peoples' (2000) 110 *Ethics* 697; F. Teson, 'Some Observations on John Rawls' The Law of Peoples' (1994) 88 *American Society of International Law Proceedings* 18.

[43] This vision elicits much criticism, some of which has been addressed in the first part of the book. However, my goal with this brief subsection is not to provide a detailed analysis of how human rights have been integrated in all different justice theories. My purpose here, instead, is to present some of the main theoretical aspects of human rights to lay the groundwork for an assessment of the international human rights system with my global distributive justice benchmarks in the remaining part of this chapter.

As with the legalist approach, some criticise the fact that the moral discourse on human rights does not seem to consider the political values at the source of the norms, considering instead that these rights emerge from the very humanness of protected individuals.[44] In fact, people in general talk more about the link between the moral and legal discourses, the latter referred to as a concrete display of the former.[45] However, there is another important factor that should inform discussions on human rights foundations: the political discourse. Indeed, politics is involved in many spheres of the human rights regime. For example we can see the influence of politics in the basic moral judgments leading to human rights content, in pre-codification negotiations, and in giving key responsibilities to existing statist entities for enforcement of and safeguarding human rights. A political analysis highlights that social movements and, above all, the forces of hegemony play a crucial practical role in the preservation of the human rights system. Indeed, looking at human rights with a political lens can help to put legal and moral values in context while emphasising the real interests and powers lying at the source of the very production and preservation of particular *truths*.[46] In doing so, power issues and dominant interests need to be identified and exposed to disturb the order already in place; this can be done by questioning some crucial elements of a system supposedly based on neutrality and universality. Getting a sense of the political discourse underlying human rights therefore appears crucial to a real and complete understanding of the institution of human rights and to avoiding the trap of the *illusion of concord* often wrongly associated with human rights. It does not appear appropriate to envision the human rights discourse as a neutral moral system with which every state and individual agrees,[47] as we shall see in more detail when I discuss human rights conceptualisation.

In the remainder of this chapter, I will ensure that these three different aspects of human rights are taken into account to provide a more complete and realistic account of the system in relation to my global distributive justice ideals.

---

[44] M. MacDonald, 'Natural Rights' in J. Waldron (ed), *Theories of Rights* (Oxford: Oxford University Press, 1984) p. 21.

[45] K. Hessler, 'Resolving Interpretive Conflicts in International Human Rights Law' (2005) 13:1 *The Journal of Political Philosophy* 29; the preambles of the two 1966 Covenants (CCPR and CSECR) also state that human rights derive from the inherent dignity of human beings.

[46] T. Evans, *US Hegemony and the Project of Universal Human Rights* (London: Macmillan, 1996).

[47] A.J. Langlois, 'Human Rights: The Globalization and Fragmentation of Moral Discourse' (2002) 28 *Rev. Int'l Stud.* 479, at 484.

4.2 DISTRIBUTION, ACCESS, JUSTICE AND THE IHR SYSTEM: AN ASSESSMENT

This second part of the chapter aims to assess the IHR system with the benchmarks of justice established in the theoretical part of the book. I will compare the ideal role that should play in terms of access and distribution of health and genetic advancements with the current reality. Even if it could be seen as having an important role to play in fostering equitable access to genetic innovation, it appears crucial to evaluate whether this system is adequately constructed to accomplish such a goal. To examine these issues, I need to assess the human rights system with the analytical approach developed earlier. My main standard to test the human rights system is universal access to some benefits of genetics to improve health while seeking the broader goal of equality of opportunities. I will begin with an evaluation of the international aspects of human rights, discussing the notion of universalism and assessing the global order as a platform for human rights development. Then, I will address the existing relationship between human rights, health-related duties, and access to health and genetics. I will conclude with a third part on human rights conceptualisation to assess the real importance of the market reality in the realisation of access to health through the application of human rights.

4.2.1 The Global/Universal Aspect of Access to Health in the Context of the IHR System

Adopting a global and international focus, we need to first study the compatibility of the IHR system with the concept of universalism to see if the system is truly geared towards a cosmopolitan ideal where every human being is considered as a unit of moral concern. To this end we will discuss universalism in relation to relativism, individualism and 'westernalisation'. We will then provide a succinct evaluation of the global order under which human rights developed.

4.2.1.1 IHR and universalism

Different critiques of the universal aspect of human rights have been brought forward. Indeed, it has been criticised for not being sensitive to cultural specificities, for encouraging harmful individualism, and for being unduly influenced and shaped by western values.

Universality vs relativism of human rights principles The universal aspect of human rights principles gives rise to important debates. While some endorse a universal position that transcends nationality, religion, and culture and from which a limited number of principles emerge, others reject such a position, arguing instead for moral and cultural relativism on the basis that the common foundation of human nature and universal community are lacking.[48]

The proponents of the latter position believe that there cannot be a single vision of right and wrong; it depends on traditions, geography, culture, and history. They think that human rights are shaped by human agents in specific contexts and that relying on one version only would be inappropriate and too constraining. Relativists believe that even when visions and values converge across cultures, such agreement is not morally meaningful but only represents some kind of coincidence. They believe that the reality of the fragmented world prevents universal acceptance of values that are often very contextual.[49] The only universality relativists are willing to accept is one that could emerge almost accidentally from a common justification of human rights coming from different people of various traditions adopting their own references. Rorty says that rights exist to 'summarise our culturally influenced intuitions about the right thing to do in various situations, … thereby heightening the sense of shared moral identity which brings us together in a moral community'.[50] This position implies that human rights are not a reality that we can take out of context and support as a universal answer, but instead are more a part of a culture endorsed by specific societies and communities.

While the language of human rights can be broad and neutral enough to allow respect for and sensitivity to various cultural and contextual differences,[51] this book focuses on the idea that these rights are grounded in the equal respect that each human being deserves. This egalitarianism arises from human agency and from some objective features, shared by everyone, which deserve universal consideration.[52] Some of those typically human characteristics refer to the capacity to think, to connect morally with others, to have a

---

48 On these two positions, see S.R. Benatar, A.S. Daar and P. Singer, *supra* chapter 2, note 1; K. Booth, *supra* note 36.

49 On this relativist position, see J. Chan, 'A Confucian Perspective on Human Rights for Contemporary China' in J. R. Bauer and D. A. Bell (eds), *The East Asian Challenge for Human Rights* (Cambridge: Cambridge University Press, 1999) p. 212.

50 R. Rorty, 'Human Rights, Rationality, and Sentimentality' in S. Shute and S. Hurley (eds), *On Human Rights: the Oxford Amnesty Lectures 1993* (New York: Basic Books, 1993) p. 117.

51 Some even say that human rights and their basic principles are echoed in every tradition of the world. For more on this view, refer to R. Coomaraswamy, 'Reinventing International Law: Womens' Rights as Human Rights in the International Community' in P. Van Ness (ed), *Debating Human Rights: Critical Essays from the United States and Asia* (London: Routledge, 1999) p. 167, at p. 169.

52 This position is shared by M. C. Davis, 'Constitutionalism and Political Culture: The Debate over Human Rights and Asian Values' (1998) 11 *Harv. Hum. Rts. J.* 109; F.R. Tesón, 'International Human Rights and Cultural Relativism' (1985) 25 *Va. J. Int'l L.* 869.

conception of the good life; to feel hunger, pain, and sexual desire; and to want our needs to be fulfilled, to judge, to dream, etc.[53] Therefore this calls for universally requiring the protection of socio-economic rights to ensure the preservation of humans' potential for self-realisation. These rights should not be defined or justified in relation to any particular legal system, state, or community. Martin is clear when he defines human rights as values that should be considered 'reasonable by persons at different times or in different cultures … principles, [that] would be thought to have connection with a fairly wide range of differing conventional moralities'.[54]

Another way to justify universal principles is to refer to shared understandings and conceptions of crucial elements related to the perception of human nature. For example, many experts from various disciplines like anthropology, sociology, philosophy, and the social sciences have demonstrated, with empirical research, that human beings often share comparable ideas of acceptable and unacceptable conduct and behaviour towards other individuals.[55] It thus appears possible to reach, through a flexible and cross-cultural exchange, some sort of agreement on universal rights we can all respect.[56] This illustrates a sense of unity transcending borders and cultures, the existence of a single moral community in certain identified spheres, and standards of social justice and human dignity, all of which support our cosmopolitan perspective. This is why relativism and particularism associated with specific states' actions and cultures should not be prioritised, as they can encourage passivity towards the shared universal concerns for human wrongs translated in international human rights.[57]

As discussed earlier, although a communitarian approach can be of great relevance in resolving numerous justice issues, I take a universal perspective on health and the role it can play in the pursuit of the good life. Indeed, health is a crucial factor in bringing people to the level where they can benefit and profit from equality of opportunity. This is why I argue that health protection should not be left to different states' voluntary initiatives, but instead to more

---

[53] B. Parekh, 'Non-Ethnocentric Universalism' in T. Dunne and N.J. Wheeler (eds), *Human Rights in Global Politics* (Cambridge: Cambridge University Press, 1999) chapter 4, p. 128, at p. 143.

[54] R. Martin, *A System of Rights* (Oxford: Clarendon, 1993) at 75; see also R.J. Vincent, *Human Rights and International Relations* (Cambridge: Cambridge University Press, 1998) who refers to human rights as a *lowest common denominator* (at p. 49).

[55] For more on this, refer to M. Nussbaum, 'Human Functioning and Social Justice: in Defence of Aristotelian Essentialism' (1992) 20 *Political Theory* 222.

[56] B. Parekh, *supra* note 53.

[57] K. Booth, *supra* note 36.

global and coordinated actions.[58] In this sense, I consider that the 'universality' aspect of the international human rights legal and moral discourse is appropriate for health and compatible with my justice framework.

Universality of rights and individualism In their inclusiveness and by granting the same rights and privileges to everyone, universal human rights are also criticised for allowing detachment from the reality in which individuals evolve. If we take this view, rights seem to be the cause of a deep sense of selfishness and individualism cultivated in modern societies. They can isolate people from one another.[59] Indeed, some argue that even if IHR are supposed to be established to ensure assistance from an entity that can facilitate individual prosperity, there are more useful ways to encourage commitment, solidarity, and accountability than human rights.[60] Those who support this vision argue against the individual focus typical of the universal human rights language because it can have the perverse effect of ignoring collective responsibilities. For example, societies where the majority of wealth is controlled by a few agents while many others have rights, but still end up with much less than what they need, are not rare. In this sense, universal human rights can be viewed as passively encouraging a social division that can prevent a true community from flourishing. This powerful quote from Wendy Brown exemplifies this argument: '[i]n the same gesture with which rights draw a circle around the individual, in the very same act with which they grant her sovereign selfhood, they turn back upon the individual all responsibility for her failures, her condition, her poverty, her madness – they privatise her situation and mystify the powers that construct, position, and buffet her'.[61] In reaction to what they call the ethnocentric and patriarchal tone of the language of rights,[62] Sen and Nussbaum adopt the concept of human beings' capabilities. Instead of granting human beings individual rights, they argue that we should provide them with an appropriate social basis to develop their capabilities. Even if they believe in the same special nature of human beings and in the fact that all should somehow benefit from equal and universal treatment, they do not believe that rights are always the appropriate tools to safeguard universal and global equality.

In response to this critique of rights, we could say that the language of human rights is not at all incompatible with the concept of human solidarity,

58 J. Mann et al. *supra* note 7, at p. 21 et seq.

59 J. Waldron, *Nonsense upon Stilts: Bentham, supra* chapter 2, note 82, part 1.

60 D. Kennedy, *supra* note 23.

61 W. Brown, *States of Injury: Power and Freedom in Late Modernity* (Princeton: Princeton University Press, 1995), chapter 5, at p. 128.

62 C. Brown, *supra* Part I, note 1.

but that it really exists to bring all people – regardless of their citizenship – to a level where they can enjoy full lives within their community and to the benefit of their fellow community members.[63] Indeed, the universal aspect of human rights by which our similarities are acknowledged encourages cohesion for sharing the benefits of rights and burdens of associated duties. In other words, rights should be seen as part of a 'reciprocal universality' which makes it impossible to see right-holding as a totally selfish and individual experience.[64] Mandela, when he refers to his country's liberation, highlights the importance of rights' universality in fostering a global sense of solidarity: '[o]ne of the striking features of modern times is the number of men and women all over the globe, in all continents, who fight violation of human rights'.[65]

Universalism vs the influence of western values A common critique of the international system of human rights is directed at the strong influence of western values and interests in the construction of this so-called universal structure. As discussed above, IHR law mainly governs relations between state governments and their citizens – highlighting, for some, a strong western influence hardly compatible with human rights' universality. Indeed, it is argued that liberal and economic theories have greatly influenced the development of the dominant human rights discourses, and that these influences embrace a philosophy where individualism prevails and in which people are seen as isolated abstractions focusing on their own interests.[66] For Donelly, the human rights system is mainly concerned with civil and political rights of citizens from liberal and democratic welfare states.[67] Some East Asian political leaders agree, arguing that the public disputes and individual pursuit of private interests typically associated with political and civil human rights are in no way universal, because they are incompatible with traditional Asian values of social harmony and community interests.[68] Indeed, many more scholars from various perspectives criticise the system on the basis that it does not reflect universal human values but instead endorses a unique, western, bourgeois,

---

63 L. Doyal and I. Gough, *supra* chapter 1, note 42, at p. 136.

64 C. Jones, *supra* chapter 1, note 10, at pp. 80–83; A. Gerwirth, 'Rights', in L.C. Becker and C. Becker (eds), *Encyclopedia of Ethics* (London: Garland, 1992) at p. 1108.

65 Cited in K Booth, *supra* note 36 at p. 57.

66 J. Mann et al., *supra* note 7, chapter 16.

67 J. Donnelly, 'Human Rights and Asian Values: A Defense of "Western Universalism"' in J.R. Bauer and D.A. Bell (eds), *The East Asian Challenge for Human Rights* (Cambridge: Cambridge University Press, 1999) p. 60, at p. 68.

68 As reported in C.R. Beitz, 'Human Rights as a Common Concern' (2001), *supra* note 42.

liberal, and masculine way of envisioning humaneness.[69] In this picture, the UN is perceived as an entity whose goal is to support and promote the liberal values enshrined in normative documents. Those liberal human rights standards are often used as universal civil and political thresholds that countries must meet to receive support from the wealthiest. However, in most cases, these norms do not play a great role in actually improving countries' socio-economic reality, but instead accept and even encourage persistent inequalities within the existing order 'where market efficiency, discipline and confidence, economic policy credibility and consistency are often awarded higher priority than issues of dignity and rights'.[70] In response to this, some instead view the western influence on human rights as very contextual, associated with a period and context that could have been completely different at other times, and that could also change in the coming years with, for example, the ascent of Asian powers.[71] The argument about the strong influence that western values have on human rights raises important concerns about the true universalism of human rights. I believe that it is important to differentiate between human rights' content and application. Indeed, even if socio-economic rights have been established in a western liberal context, the values they defend – for example rights to health, nutrition, and work – can remain universal. In my view, the strong western influence has a negative impact on human rights universalism when the time comes to implement and enforce those rights and nothing is done to reassure excluded groups about the protectiveness and inclusiveness of human rights.[72] I will come back to the dominant political discourse that lies behind the realisation of human rights when I address the conceptualisation of human rights within the reality of the market.

This last subsection demonstrates that the universal quality of human rights can be questioned on many fronts. Indeed, although the IHR system appears to be based on the protection of individual interest, it can foster different outcomes, depending on how it is interpreted and circumscribed. In order to get a better idea of whether and how each individual's interests get taken into consideration through the application of IHR, we need to say a few words about the global order in which they develop.

---

69 For example, see R. McCorquodale and R. Fairbrother, 'Globalization and Human Rights' (1999) 21:3 *Human Rights Quarterly* 735, at 740.

70 T. Evans, 'Universal Human Rights', *supra* note 30, at 166; M. Mutua, *Human Rights: A Political and Cultural Critique* (Philadelphia: University of Pennsylvania Press, 2002) at p. 35 and chapter 4.

71 K. Booth, *supra* note 36, at pp. 52–53.

72 T. Evans, *supra* note 30, at 166.

4.2.1.2 Assessment of the global order under which IHR develops

Article 28 of the UDHR says that '[e]veryone is entitled to a social and international order in which the rights and Freedoms set forth in this Declaration can be fully realised'. As Pogge clearly explains, this article does not add new human rights to those already existing, but serves to establish that human rights are 'claims on the institutional order of any comprehensive social system',[73] and that institutional orders should be evaluated in relation to the impact they have on the realisation of human rights.

States are at the centre of the human rights system. Even if, as we just saw, a strong argument can be made to support human rights' universality, granting, implementing, and enforcing rights remains the first responsibility of states in international law.[74] This can be problematic in that state sovereignty can conflict directly with the universality principle. Indeed, one important rule in international law is that states are sovereign entities and are entitled to set up their own rules and norms within their territorial borders.[75] However, states can cede part of their sovereignty voluntarily in agreeing to comply with international standards like IHR. When they do, they become accountable for their actions in that sphere. In reality, however, things are not that simple; affluent states are typically reluctant to concede any of their sovereignty to supranational institutions.[76] Sovereignty remains an essential principle of international law and plays a crucial role in how states behave and interact with one another. The exercise of one's human rights is directly linked with the nature of existing national legislative and institutional mechanisms in place. This is especially relevant as the reference to IHR enforcement mechanisms is conditional on prior exhaustion of all national remedies.[77] This also means that, contrary to what universalism requires in terms of equal consideration for all human beings' interests, people can best exercise their rights as citizens as opposed to human beings.[78]

However, the world has become so interrelated in every sphere of activity that individuals are increasingly linked to each other through different modes of interaction and dependence. As explained by Monshipouri et al., 'what has happened through conditions of chronic globalisation is that the fate of

---

73 T.W. Pogge, 'Human Rights and Global Health: A Research Program' (January 2005) 36:1/2 *Metaphilosophy* 182, at 196.

74 M. Koskenniemi, 'The Future of Statehood' (1991) 32 *Harv. J. Int'l L.* 397.

75 This principle is codified in the Charter of the United Nations, 892 *U.N.T.S.* 119, art. 2(1).

76 E. O'Keefe and A. Scott-Samuel, 'Human Rights and Wrongs: Could Health Impact Assessment Help?' (Winter 2002) 30:4 *The Journal of Law Medicine & Ethics* 734.

77 J. Mann et al., *supra* note 7, chapter 2.

78 C.R. Beitz, 'Human Rights as a Common Concern', *supra* note 42 at 274.

communities throughout the world has become linked through complex and dynamic systems that create moral connections between the agents and the subjects of social action regardless of territorial and political boundaries'.[79] In such a context, human rights are meant to assert universal claims that people can have to resources and also to the protection of their inherent dignity as actors in this global reality.[80] This is compatible with the recognition, in international law, of the role of states in the protection of human rights – not only inside but also outside of their national borders, for the benefit of non-citizens.[81] Indeed, when states agree to UN membership, they commit to 'achieve international cooperation in solving international problems'.[82] Also, when the ICESCR requires states to take all necessary action within their means to achieve the full realisation of protected rights, it really asks that they do so within their budgetary capacities but also with the help of technical assistance and international cooperation.[83] In other words, the ICESCR clearly provides a normative foundation for state obligations to foreigners located outside their territories, in part through their external trade and cooperation activities. The role of international cooperation in human rights enforcement has been highlighted by the former UN High Commissioner for Human Rights when she refers to protection of the right to health, saying, '[s]pecifically, State parties should recognise the essential role of international cooperation and comply with their commitment to take joint and separate action for the full realisation of the right to health, taking into account the gross inequality in the health status of people, particularly between developed and developing countries'.[84]

The requirement to meet such extraterritorial obligations can infringe on state sovereignty (both on the giving and receiving end), and achieving a balance between human rights protection and respect for state sovereignty can

---

79 M. Monshipouri, C.E. Welch and E.T. Kennedy, 'Multinational Corporations and the Ethics of Global Responsibility: Problems and Possibilities' (2003) 25:4 *Human Rights Quarterly* 965, at 969.

80 J. Mann et al., *supra* note 7, chapter 16.

81 For a clear exposé of this topic, especially on the link between states' obligations and the activities they undertake across borders, refer to S. Skogly and M. Gibney, 'Transnational Human Rights Obligations' (2002) 24:3 *Human Rights Quarterly* 781.

82 UN Charter, art.1 (3), signed 26 June 1945, 59 Stat. 1031, T.S. No. 993, 3 Bevans 1153 (entered into force 24 October 1945).

83 ICSECR, art. 2 and 23; Committee on Economic, Social and Cultural Rights, *General Comment No 14: The Right to the Highest Attainable Standard of Health, Article 12 of the Covenant,* 12 May 2000, E/C. 13/2000/4, online on the website of the UNHCHR: http://www.unhchr.ch/tbs/doc.nsf/(symbol)/E.C.12.2000.4.En?Open Document (accessed 30 March 2009), para. 75.

84 UN Economic and Social Council, 'Report of the High Commissioner: The impact of the Agreement on Trade-Related Aspects of Intellectual Property Rights on Human Rights', 27 June 2001, E/CN.4/Sub.2/2001/13, para. 35

be somewhat challenging in practice. In that sense, international human rights can contribute to ending the unlimited sovereignty that states traditionally have over their people's entitlements.[85] Those changes in state sovereignty are not properly addressed by human rights. Indeed, although the global community has changed significantly over the last 50 years – with more members, more diversity, new powerful agents, closer ties and, at the same time, greater divisions – IHR have not managed to adapt to these changes.[86] For example, international human rights do not provide remedy for violations committed by non-state actors like transnational corporations, asking states to take on responsibility for what is happening within their territories and focusing on what they can do to improve their people's well-being. Doing so, the language of IHR does not acknowledge that states' control, freedom, management abilities, and flexibilities are, in reality, eroding to the benefit of non-state actors who are shaping the global society.

This changing role of states due to globalisation can have important effects on the realisation of true universal human rights. Indeed, more and more, states must respond to market forces and act to support the broader global order, which favours freedom of production and appropriation. Consequently, the global protection of IHR standards loses its universal character and becomes a tool to further economic ends rather than being an end in itself.[87] In other words, some powerful external agents involved in changing the face of international relations and whose actions are not easily controlled by national or international agencies can end up with much control over the way human rights are realised within the global order.[88] This explains, at least in part, why most socio-economic rights have not been considered seriously, despite the fact that the universal character of human rights calls for their inclusion in the emerging global consensus. This means that state sovereignty is transforming to serve different, more powerful interests, a situation that brings Chimni to say that '[a]s things stand now, the neo-colonial third world states will continue to exist but essentially in the service of the TCC [transnational capitalist class] and the global state. The Northern/Western state, on the

---

85 On the topic of human rights globalisation, see *Vienna Declaration and Programme of Action*, U.N. Doc. A/CONF.157/23, 1993, part I, art. 4.

86 A.J. Langlois, 'Human Rights: the Globalisation and Fragmentation of Moral Discourse', *supra* note 47.

87 R.W. Cox, 'Civil Society at the Turn of the Millennium: Prospects for an Alternative World Order' (1999) 25 *Rev. Int'l Stud.* 3; T. Evans, 'International Human Rights Law as Power/Knowledge' (2005) 27:3 *Human Rights Quarterly* 1046.

88 M.K. Addo, *Human Rights Standards and the Responsibility of Transnational Corporations* (Boston: Kluwer Law International, 1999); P. Alston, 'The Myopia of the Handmaidens: International Lawyers and Globalization' (1997) 8 *Eur. J. Int'l L.* 435.

other hand, will continue to shape the form and content of the emerging global state to realise TCC interests'.[89]

Another problem identified with the existing global order is that it seems to support institutions involved in human rights violations. In fact, international economic organisations with effective enforcement powers – like the WTO, the Bretton Woods Institutions (including the World Bank and the International Monetary Fund), as well as transnational corporations and other central features of the present global order – are often described as institutions of artificial global unity that systematically contribute to the persistence of severe poverty.[90] For example, although the World Bank has undertaken over 600 judicial reform projects aimed at improving the condition of the least well off, it seems to be part of a larger effort to facilitate transactions, protect property rights, and establish a stable investment environment in priority.[91] Some respond to this criticism by saying that the international order and its economic institutions are just since they give an equal chance to sovereign states to bargain and negotiate with each other and consent to the outcome. This argument ignores the fact that parties do not have the same economic bargaining power[92] and that the weakest countries almost always have to make concessions that go against their basic interest.[93]

Many suggestions have been put forward to transform the current world order and make it less burdensome for the less affluent. Some have proposed to impose the same human rights standards on the activities of non-state actors – especially transnational corporations and international economic organisations – to support a more humane globalisation.[94] It actually seems that unless we proceed with this shift, the human rights dialogue will be designed in a way to disturb those powerful actors' activities as little as possible. Going one step

---

89 B.S. Chimni, 'International Institutions Today: An Imperial Global State In The Making' (February 2004) 15 *Eur. J. Int'l L.* 1, at 6.

90 For example it has been reported that the OECD *Convention on Combating Bribery of Foreign Public Officials in International Business Transactions* is ineffective in preventing bribery by companies.

91 K. Rittich, 'The Future of Law and Development: Second Generation Reforms and the Incorporation of the Social' (Fall 2004) 26 *Mich. J. Int'l L.* 199, at 217.

92 According to the World Bank's statistics, developed countries represent 15.5% of the world's population while controlling 80.4% of the world's income. World Bank, *World Development Report 2005* (New York: Oxford University Press, 2005) at 257.

93 A. Anghie, 'Time Present and Time Past: Globalization, International Financial Institutions, and the Third World' (2000) 32 *N.Y.U. J. Int'l L. & Pol.* 243, at 274.

94 S. Picciotto, 'Democratizing Globalism' in D. Drache, *The Market or the Public Domain, Global Governance and the Asymmetry of Power* (New York: Routledge, 2001) p. 335.

further, we also have to question the real nature of the human rights standards whose application we want to extend. Indeed, as briefly mentioned above, today's human rights are viewed by some as inspired by cultural imperialism, and more specifically by western liberal values linked with the global political economy and aimed at protecting liberal freedoms to ensure the satisfaction of private interests, especially those lying in property rights.[95] The content and the true importance of socio-economic rights need to be reaffirmed and strongly enforced. However, many obstacles stand in the way of such an exercise.

This section has assessed one aspect of human rights: the universal perspective of the IHR discourse with a cosmopolitan standard of global access. My analysis of the global aspect of access to health and human rights highlights a progressive weakening of the cosmopolitan approach to solidarity that should be fostered by the very institution of IHR. I have realised that the concept of universalism, crucial to justifying global distributive justice and access in health, is a morally contested notion in the field of IHR, and also not always supported by the institutional structures and the politics of human rights. Indeed, the capitalist and undemocratic features of the global order seem to reduce the potential for real and universal global distributive justice supported by IHR principles. This is illustrated by the fact that the most obvious economic disparities and injustices remain unnoticed and unpunished by international law, and that the most vulnerable groups do not receive the protection they deserve under the legal system. In other words, instead of fostering a universal application of human rights, economic globalisation supports some groups, interests, and rights over others.[96] It therefore does not recognise, in practice, the universal importance of health for every human being. This problem is rarely addressed under the dominant legalist voice of the human rights discourse, which can be seen as eluding the real power relations lying at the core of many human rights violations. Because of the importance of this dominant discourse, most people ignore how human rights ideals have been transformed and modelled by external powers that often bring them, wrongly, to believe in a just world.[97] This is why it is crucial to undertake a deeper analysis of the complex global political reality within which the system of human rights evolves. I will come back to this in the third part of this section, when we discuss the conceptualisation of human rights within the reality of the market.

---

[95] K. Marx, 'On the Jewish Question', in J. O'Walley (ed), *Early Political Writing* (Cambridge: Cambridge University Press, 1994).

[96] A. Hoogvelt, *Globalization and the Postcolonial World: The New Political Economy of Development* (Basingstoke: Palgrave, 2001).

[97] S.R. Benatar, 'A Perspective from Africa on Human Rights and Genetic Engineering' in R. Dawkins (ed), *The Genetic Revolution and Human Rights* (Oxford: Oxford University Press, 1999).

Before doing so, I need to evaluate whether this system is compatible with other benchmarks of my framework of distributive justice in health.

4.2.2 Legal Access to Health, Responsibility for Distribution of Health and Human Rights

The main goal of this subsection is to determine how the need for health improvement through genetic access materialises in legal rights, entitlements, and related responsibilities for distribution, and whether the actual human rights system is working toward this goal. I will start with an assessment of socio-economic rights, and more specifically, health-related rights with benchmarks of distributive justice. In the second part, I will address the integration of these rights with the associated responsibility to distribute genetic technologies to promote real access to health improvement and equality of opportunities, within the mechanism of human rights implementation.

4.2.2.1 Envisioning socio-economic rights with benchmarks of distributive justice

Rights can be crucial in a theory of justice as they can impose restrictions on actions (civil and political rights) and obligations to undertake other actions (socio-economic rights). Brown states that 'the language of rights has become the way in which humanitarian impulses are expressed in the modern international system'.[98] If we take human rights as an important element of our global distributive justice theory, they will accordingly impose restrictions on distributive arrangements (social, political and economic) supported by the global order, and serve as a basis to rectify injustices created by unequal distribution. In this sense, just distribution will be achieved when individuals obtain what they are entitled to, by right, in terms of resources and opportunities. As briefly explained at the beginning of this chapter, my main focus is on the content, conceptualisation, and realisation of socio-economic rights. These rights represent claims to social equality and refer to prospects and circumstances that can allow individuals to live as actors and enjoy a good standard of living.[99] They are critically important with respect to health issues, to ensure that each individual has equal access to appropriate health care, technology, and resources to have a normal range of opportunities in other spheres of activities, as well as to be able to take advantage of other civil and political rights. One major issue with the enforceability of economic, social, and

---

98 C. Brown, *supra* Part 1, note 1, at 103.

99 G.A. Mower, *International Cooperation for Social Justice: Global and Regional Protection of Economic/Social Rights* (Westport: Greenwood Press, 1985) at p. 3.

cultural rights is the confusion between the recognition of rights in themselves and the degrees of realisation of those rights in terms of implementation and protection. Indeed, as mentioned previously, the realisation of health-related rights is dependent on legal interpretation, decisions on resources allocation, and political convictions.[100]

Definition of the right to health In the last few decades, human rights have been associated with the crucial goal of achieving acceptable standards of health.[101] Rights emerge from our general theory of global distributive justice in health. It gives us indications as to which kinds of claims should be viewed as rights, which needs should be codified as rights, and against which standards *just* social rules, institutions, and people who establish and support them should be assessed. Rights are grounded in the basic moral interests that individuals have in their content. Some associate basic rights-protection with primary necessities and the preservation of human life. Such basic rights can emerge from the basic needs shared by every human being – such as subsistence, water, housing and health care[102] – and they can be ordered according to the nature of the interest they aim to protect. Shue establishes some priorities among rights, putting the fulfilment of basic rights first,[103] followed by non-basic rights, culture enrichment, and, finally, mere satisfaction of preferences.[104] Meeting basic rights in priority is crucial in that it helps free people from oppression and discrimination, removes a certain degree of vulnerability that exposes them to the power of others, and allows them to exercise many other rights.[105] Another way to describe it is to refer to the notions of 'social citizenship', 'personhood' or 'moral agency', which justify demands for access to basic necessities for survival and potential for a good life.[106] This relates to my argument for a cosmopolitan approach to genetic access and distribution; the superior basic human interest in health is shared by everyone, it represents an appropriate focus for a duty not to harm and ensures that individuals are in a position to profit from equality of opportunities, plan for a

---

100 J.K. Mapulanga-Hulston, 'Examining the Justiciability of Economic, Social and Cultural Rights' (Winter 2002) 6:4 *The International Journal of Human Rights* 29, at 42–43.

101 For example, refer to V.A. Leary, 'The Right to Health in International Human Rights Law' (1994) 1 *Health and Human Rights* 24.

102 P. Jones, *Rights* (London: MacMillan, 1994) at pp. 13–15.

103 Basic rights are those grounded in basic human needs and establishing the threshold under which no one should be entitled to fall to avoid degrading inequalities.

104 H. Shue, *Basic Rights*, *supra* chapter 2, note 63, chapter 5, at p. 111.

105 P.D. Jacobson and S. Soliman, 'Co-opting the Health and Human Rights Movement' (Winter 2002) 30:4 *The Journal of Law, Medicine & Ethics* 705.

106 S.R. Benatar, A.S. Daar and P. Singer, *supra* chapter 2, note 1.

good life, and pursue their goals. When we use the language of socio-economic human rights in this specific context, it can refer to the right to health (art. 12 ICESCR).[107] For the need to assess tangible genetic access and distributive justice, it indeed appears appropriate to focus on the right to health, which is a controversial and inclusive right that can cover a wide range of activities, products, and technologies. Nevertheless, much of what I am saying about the human right to health, aside from the content and definition, could be used in an analysis of other socio-economic rights.

The right to health, including health care, services, and technologies, is accepted as a human right in regional and international law.[108] Indeed, it has been protected by the WHO Constitution since 1946[109] and was further enshrined in numerous other human rights treaties.[110] The ICESCR offers the strongest and most ambitious version of the right to health in its article 12:

> 1. The States Parties to the present Covenant recognise the right of everyone to the enjoyment of the highest attainable standard of physical and mental health.
> 2. The steps to be taken by the States Parties to the present Covenant to achieve the full realisation of this right shall include those necessary for:
>   (a) The provision for the reduction of the stillbirth-rate and of infant mortality and for the healthy development of the child;
>   (b) The improvement of all aspects of environmental and industrial hygiene;
>   (c) The prevention, treatment and control of epidemic, endemic, occupational and other diseases;
>   (d) The creation of conditions which would assure to all medical service and medical attention in the event of sickness.

---

[107] It can also refer to the right to enjoy the benefits of scientific progress and its applications (art. 15 (1)(b) ICESCR).

[108] For more on this definition, see T. Goodman, 'Is there a right to health?' (2005) 30 *Journal of Medicine and Philosophy* 643.

[109] *Constitution of the World Health Organization*, adopted by the International Health Conference, New York, signed on 22 July 1946 by the representatives of 61 States.

[110] For example, the right to health has been codified in: *Universal Declaration of Human Rights*, art. 25; *International Covenant on Economic, Social and Cultural Rights*, art.12; *Convention on the Elimination of all Forms of Discrimination against Women*, G.A. res. 34/180, 34 U.N. GAOR Supp. (No. 46) at 193, U.N. Doc. A/34/46, entered into force 3 September 1981, art.12; *Convention on the Rights of the Child*, G.A. res. 44/25, annex, 44 U.N. GAOR Supp. (No. 49) at 167, U.N. Doc. A/44/49 (1989), entered into force 2 September 1990, art. 24; See also, *African Charter on Human and People's Rights*, adopted 27 June 1981, OAU Doc. CAB/LEG/67/3 rev. 5, 21 I.L.M. 58 (1982), entered into force 21 October 1986, art. 16; *Additional Protocol to the American Convention on Human Rights in the Area of Economic, Social and Cultural Rights*, 1988, art. 10.1.

The Committee on Economic, Social and Cultural Rights interprets the right to health broadly as including the right to facilities, goods, services, education, and research required to achieve the highest attainable standard of health.[111] Nevertheless, although the right to health is codified in many legal instruments and is meant to be interpreted broadly, it nevertheless lacks 'conceptual clarity'.[112]

Despite these lacunae, there seems to be some agreement on some of the essential elements that the right to health should include, irrespective of state resources and of individuals' economic situation. These elements can be divided into two categories, the first related directly to health care (including, for example, access to basic medical treatment and services for severe diseases, maternal and child care, and immunisation against infectious diseases) and the other related to determinants of health (such as access to education and prevention methods, drinkable water, food, and adequate sanitation).[113] To ensure that the right to health can be respected as much as possible, another important aspect of its content relates to the international obligations associated with preventing violations of the right, facilitating access to health-related products and services, and providing aid to other countries.[114] As discussed at the very beginning of the book, the science of genetics has the potential to be used and developed to help address and resolve most of the essential elements included in the right to health.

Nonetheless, what often happens in reality is that, when referring to health rights, people only include basic rights to emergency health care and subsistence, indirectly endorsing the fact that access to adequate and adapted health care remains reserved for a privileged few.[115] This is in line with the application of Rawls' difference principle. This position is unsatisfactory since the scope of the underlying elements of the difference principle remains very

---

111 A.R. Chapman, 'The Human Rights Implications of Intellectual Property Protection' (2002) 5 *Journal of Int'l Eco. L.* 861 at 880.

112 A.E. Yamin, 'Protecting and Promoting the Right to Health in Latin America: Selected Experience from the Field' (2000) 5:1 *Health and Human Rights* 117; B. Toebes, 'Towards an Improved Understanding of the International Right to Health', *supra* note 10, at 662.

113 *Declaration of Alma-Ata*, 12 September 1978, reprinted in Report of the International Conference on Primary Health Care (Geneva: World Health Organization, 1978), art.VII; Committee on ESCR, *General Comment No 14*, *supra* note 83, para. 4, 11 and 12; B. Toebes, 'The Right to Health' in A. Eide et al. (eds), *Economic, Social and Cultural Rights: A Texbook*, 2nd edn, (London: Maartinus Nijhoff, 2001), chapter 10, p. 169.

114 Committee on ESCR, *General Comment No 14*, *supra* note 83, paras. 39 and 45.

115 S. Caney, *supra* chapter 1, note 60.

unclear. It gives no precise indication as to what criteria must be met to be considered as belonging to the least privileged group. Also, it does not go into detail about what is required to reach the *greatest benefit of the least advantaged group* threshold, as it does not impose a floor below which no one should be allowed to fall. We argue that focusing on the most urgent health needs of any individual or on the category of the absolute poorest people is far from enough, as many individuals can still be left in need of the benefits of genetics outside of the distribution realm. It also does not guarantee that people's basic needs will be met so that they are in a position to enjoy other rights and create a good life. Indeed, respecting rights to health, health care, and genetic technologies (as I consider that they could become the new standard of care) is essential to the broader goal of achieving equality of opportunities for all. In fact, as explained earlier, access to health in terms of availability and affordability of genetic products and services will allow people to seize opportunities towards achieving rewarding lives, and this should be the ideal threshold for distribution. However, even if true equality of opportunities remains out of reach in a world where there are limited health resources for unlimited health needs, we believe that the interpretation of the right to health should nevertheless be widened so that it does not only refer to the most urgent needs or the absolute worst-off, but instead aims at broader and more inclusive thresholds of health, such as, for example, basic health-needs satisfaction in light of existing medical and genetic technology.[116]

Efficacy of the right to health in terms of justice In addition to issues relating to the content and validity of human rights, there is also the problem of true efficacy of international socio-economic rights and of the right to health more specifically. Many question the real justiciability of socio-economic rights because of their prevailing political character, the vagueness of the scope and content of the rights, the associated obligations of conduct (not of result), and weak supervision and compliance mechanisms.[117] Indeed, very few countries monitor and gather data on their realisation of the right to health.[118]

The *Limburg Principles* and the *Maastricht Guidelines* state that economic, social, and cultural rights violations can arise following both acts of commission and acts of omission at the national and regional levels.[119] For instance,

116 L. Doyal and I. Gough, *supra* chapter 1, note 42, at pp. 130–135.

117 A. Eide, 'Future Protection of Economic and Social Rights in Europe' in A. Bloed et al. (eds), *Monitoring Human Rights in Europe: Comparing International Procedure and Mechanisms* (Boston: Maartinus Nijhoff, 1993) p. 187.

118 A.R. Chapman, *supra* note 111.

119 *The Limburg Principles on the Implementation of the International Covenant*

member states can be brought before the European Committee of Social Rights (ECSR) to respond to alleged violations of the European Social Charter (ESC) dispositions.[120] On the international scene, the Committee on ESCR is in charge of revising periodic reports submitted by State parties to monitor implementation of socio-economic rights. To conduct its evaluation and analysis of how the right to health is respected by member states, the CESCR refers to four broad categories: general issues (proportion of countries' GNP dedicated to health, public/private standards); healthcare (provision in rural and urban regions, availability and affordability of health care and services); determinants of health (access to food, water, and sanitation); consideration of more vulnerable groups (indigenous population, HIV-infected individuals and communities).[121] Also, in its 2000 general comment on the right to health, the Committee established four different criteria to evaluate the achievement of the right to health, namely, availability, accessibility, acceptability, and quality.[122] Originally, the Committee had no mandate to review individual complaints concerning violations of economic, social, and cultural rights. However, article 2 of the last available version of the draft Optional Protocol opens the door to such complaint and stipulates that 'Communications may be submitted by or on behalf of individuals or groups of individuals, within the jurisdiction of a State Party, claiming to be [direct] victims of a [significant] violation of any of the rights set forth in the Covenant by that State Party'.[123]

---

on Economic, Social and Cultural Rights (1987) 9 *Human Rights Quarterly* 122, UN doc E/CN.4/1987/17, at 131; *Maastricht Guidelines on Violations of Economic, Social and Cultural Rights*, *supra* note 12.

120 For example, in the case *International Federation of Human Rights Leagues (FIDH) v France* (complaint no. 14/2003), the ECSR was asked to interpret art. 13(1) of the ESC on the right to medical and social assistance in relation to the *French Finance Amendment Act*. The complaint related to access to medical care by illegal immigrants with very low income. The claimant, the *International Federation on Human Rights Leagues* (FIDH) argued that illegal immigrant status should not, in any case, justify denying medical assistance to individuals, and that ensuring free treatment in case of emergencies and life-threatening conditions was not sufficient. The French Government, on the other hand, submitted that illegal immigrants did not fall within the scope of the charter's protected individuals. The ECSR admitted that the practical applicability and definition of health emergencies and life-threatening conditions were vague, but since some form of medical assistance, even minimal, was in place, it was enough for the Committee to conclude that the French Act did not violate art. 13 of the ESC.

121 UN Committee on ESCR, *Guidelines on Reporting, Questions relating to art. 12 of the CESCR*. UN doc. E/1991/23, at 88–110; B. Toebes, 'Towards an Improved Understanding of the International Right to Health', *supra* note 10, 666–667.

122 Committee on ESCR, *General Comment No 14*, *supra* note 83.

123 Revised Draft Optional Protocol to the International Covenant on Economic, Social and Cultural Rights, (24 December 2007) A/HRC/8/WG.4/2http://

My principles of distributive justice demand that, in the course of distributive endeavours, I consider the health needs of every individual as critically important and consider them as rights to the extent necessary to ensure that they have the capacity to take advantage of available opportunities. Some, like Kennedy, believe that rights are not the best instrument to help us achieve equitable distributive endeavours. In fact, he says that the very legal nature of rights does not allow us to prioritise and ensure equitable distribution among different right-holders – some more deprived, with greater and more urgent needs than others.[124] Instead, he argues, it too often allows institutions to protect some people's preferences even when others are unable to enjoy any rights.

Others, like Cullet, believe that although they are universal and awarded to all, human rights are meant to focus on the most deprived individuals and communities.[125] Indeed, the Committee on ESCR, in its analysis, seems to prioritise economically deprived people's claims over the states' limited resources.[126] This is also the position that the European Court of Human Rights adopted in the case *Airey v Ireland* in 1979.[127] Indeed, the Court interpreted the right to a fair trial very broadly, so as to include the right to civil legal aid protection. In so doing, the Court gave this right some teeth and offered tangible support for the equal treatment of everyone, with a special focus on the least affluent individuals and communities. As Scott notes, this decision illustrates that 'human rights protection can, and should, be a result of a contextual interpretive analysis of what is needed to make a right truly a right of "everyone" '.[128] Applying this decision to the right to health allows us to argue for the provision of a sufficient and adequate amount of health care to secure equality of opportunities, regardless of socio-economic factors.

This brings us to the realisation that, depending on how they are implemented, human rights can impact the health of both individuals and communities, replacing our strictly medical perspective with a broader 'social good' vision of health. Indeed, the rights to health, health care, and related technolo-

---

www2.ohchr.org/english/issues/escr/documents_5.htm (accessed 20 May 2009). Economic and Social Council, Commission on Human Rights, *Report of the Open-Ended Working Group to Consider Options Regarding the Elaboration of an Optional Protocol to the International Covenant on Economic, Social And Cultural Rights on its First Session*, E/CN.4/2004/44, 15 March 2004.

124 D. Kennedy, *supra* note 23.

125 P. Cullet, *supra* chapter 3, note 101.

126 C. Scott, 'Reaching Beyond (Without Abandoning) the Category of Economic, Social and Cultural Rights' (1999) 21:3 *Human Rights Quarterly* 633.

127 *Airey v Ireland*, 32 Eur. Ct. H.R. (ser. A) (1979), reprinted in 2 Eur. H.R. Rep. 305 (1979).

128 C. Scott, *supra* note 126, at 641.

gies, resources, and services are inherently valuable in their goal of protecting individual welfare, basic needs, and interests. In itself, it justifies their recognition as a normative foundation for justice in a just global structure. However, realising these rights necessarily entails correlative duties by identified duty-bearers against whom rights are claimed. Indeed, as rights can found certain socially-guaranteed claims, they can also provide a basis for related duties of states and other members of the global community. This second section seeks to address the relationship between human rights, correlated duties, and access to health.

4.2.2.2 Implementing the right to health

A very important aspect of the right to health, which is too often forgotten, concerns duties involved in creating the conditions necessary for ensuring practical right enforcement and fulfilment resulting in efficient and just distribution. Indeed, our very capacity to benefit from rights is directly linked to our acceptance of responsibilities.[129] On the importance of related duties, Shue says that '[i]t is only because rights may lead to demands and not something weaker that having rights is tied as closely as it is to human dignity'.[130]

Notion of duty Focusing on duties can be helpful at many levels for realising global distributive justice in health. It can encourage a dialogue on who has to do what, in which priority order, for the realisation of the right to health. It can also help to highlight some problems with the political and economic context within which human rights are conceptualised.[131]

Vincent states that rights are composed of five essentials: namely the subject and the object of the right, the way to exercise the right, the duty-bearer and the justification of the right.[132] Discussions on human rights should therefore not only cover the essence of what should be granted, but should also deal with what is necessary to achieve these rights and by whom they should be respected. This is the focus of this section.

The UDHR and the ICESCR implicitly refer to the obligation of states to secure a right to health for their people and for foreigners through international cooperation, without any discrimination. These obligations are demanding;

---

129 S. Benatar, A.S. Daar and P. Singer, *supra* chapter 2, note 1.

130 H. Shue, *Basic Rights*, *supra* chapter 2, note 63, at p. 14.

131 A. Chapman, 'Reintegrating Rights and Responsibilities' in K.W. Hunter and T.C. Mack (eds), *International Rights and Responsibilities for the Future* (Westport, CT: Praeger, 1996) p. 3.

132 R.J. Vincent, *Human Rights and International Relations* (Cambridge: Cambridge University Press, 1986) at p. 8.

they require 'action to create freedom'.[133] Other soft-law initiatives have also been undertaken to establish and propose specific duties for states in relation to socio-economic rights to secure a minimum quality of life for individuals and an adequate environment for future generations.[134] Duties are assigned both to countries and to the international community, which needs to take global inequalities between countries into account in the realisation of the right to health.[135] Indeed, human rights cannot be respected in less affluent countries if the rich countries do not respect their related duty to refrain from adopting detrimental political, military, and economic strategies against them. However, since practical implementation is almost uniquely national, international supervision is very limited, as we will see in the next section.

The right to health can give rise to three main types of obligation: the obligation to respect, protect, and fulfill (ensure and promote) individual and community health needs within and across borders.[136] These, in turn, can be divided into two broad categories: positive duties (to perform actions toward equal access to quality health care and against interference with this right; to assist; to promote health and disease prevention through legislative and policy mechanisms) and negative duties (to refrain from performing certain health-harming actions and from establishing institutions that could undermine individuals' right to health).[137]

Positive duties Positive duties to fulfil the right to health imply undertaking actions to secure sufficient amounts of goods and services to meet individuals' basic health needs with the help of existing medical and genetic technology. Negative duties involve refusing to endorse an institutional order that entails avoidable and foreseeable violation of those rights.[138] Proponents of positive duties believe that refraining from institutionally denying and undermining

---

133 E. O'Keefe and A. Scott-Samuel, *supra* note 76.

134 *The Limburg Principles on the Implementation of the International Covenant on Economic, Social and Cultural Rights*, *supra* note 119; *Trieste Declaration of Human Duties: A Code of Ethics and Shared Responsibilities*, 1995; *Universal Declaration of Human Responsibilities*, September 1997.

135 UN Economic and Social Council, *Report of the High Commissioner: The Impact of the Agreement on Trade-Related Aspects of Intellectual Property Rights on Human Rights*, 27 June 2001, E/CN.4/Sub.2/2001/13, para. 35, online on the UNHCHR website: http://www.unhchr.ch/Huridocda/Huridoca.nsf/e06a5300f90fa0238025668700518ca4/590516104e92e87bc1256aa8004a8191/$FILE/G0114345.pdf (accessed 4 March 2009).

136 H. Shue, *Basic Rights supra* chapter 2, note 63.

137 J. Narveson, *The Libertarian Idea* (Philadelphia: Temple University Press, 1988) at 57; R. Cruft, 'Human Rights and Positive Duties', *supra* chapter 2, note 90, at 30.

138 T.W. Pogge, 'Severe Poverty as a Violation of Negative Duties', *supra* chapter 2, note 91, at 66–68.

access to health goods and services will not always be enough to make a difference in helping the most vulnerable. Indeed, their view is that maintaining fair institutions does not necessarily ensure that severely disabled individuals will be able to get what they should be entitled to from those institutions. They also believe that basic needs and interests are so important that they should give rise to both institutionally grounded positive assistance duties and duties of non-interference.[139]

Fulfilling positive duties to aid raises a number of questions with regard to availability and allocation of scarce resources across time and space, imposition of budgetary priorities, income distribution, public policy making, and legislative and judiciary powers.[140] Indeed, the limited resources (human, budgetary and so on) of states, the numerous unfulfilled basic health needs, and the difficulty of finding accountable actors and institutions can raise scepticism about the very existence of a right to health and thus can be seen as obstacles to the positive implementation and enforcement of welfare claims.[141] Nervertheless, universal human rights, such as the right to health, should be appraised from a global and long-term perspective, and should be awarded the most resources and attention in comparison to preferences and other non-basic rights. To understand what this means in practice, the Committee on Economic Social and Cultural Rights issued a statement in 2007 to give more details on the scope of a state's obligations towards achieving the full realisation of the covenant's socio-economic rights 'to the maximum of its available resources'.[142]

As discussed, health is of universally great importance. Protecting health is an essential part of the duty not to harm, as it aims to bring individuals to a situation in which they are able to function, seize opportunities, and make the most out of them. Ensuring a certain level of good health for all helps avoid the persistence of degrading inequalities.[143] In this sense, as we explained in our critique of Rawls' difference principle, allowing some individuals to focus on their own preferences while denying positive duties to fulfil others' most basic needs is not morally justifiable.

---

139 R. Cruft, 'Human Rights and Positive Duties', *supra* chapter 2, note 90, at 35–37.

140 M. Jackman, 'The Protection of Welfare Rights under the Charter' (1988) 20 *Ottawa L. Rev.* 257.

141 R. Nozick, *supra* chapter 1, note 6; C. Wellman, *An Approach to Rights: Studies in the Philosophy of Law and Morals* (Boston: Kluwer Publishers, 1997) at pp. 112–114.

142 It explains that states, even when they have very limited resources, should always do everything they can to ensure the best possible enjoyment of socio economic rights to the most disadvantaged members of their society.

143 H. Shue, *Basic Rights*, *supra* chapter 2, note 63, chapter 5, at p. 111.

Although the broad language of the Covenant does not clearly establish specific actions states must take to fulfil the right to health as a means to ensure equality of opportunities, this right can give rise to different positive duties. For example, the right to health can imply a duty to establish mechanisms to prevent deprivation and encourage the provision of preventive and therapeutic health products by third parties (regulating selling costs, establishing incentives to encourage the development of health products and services for specific needs and so on). It can also include a duty to provide available health products and services to those in need with the transfer of resources at affordable cost and it can also require states to review their research priorities to take objectively serious health needs of poorer countries into consideration.[144] This latter duty to aid is often critical as it arises after some have failed in their duty to protect and to avoid harm. Many indicators can be used to assess the fulfilment of positive duties vis-à-vis the right to health. Robertson suggests five types of resources relevant to measuring human rights compliance: human, technological, informational, natural, and financial.[145] States are thus free to decide which resources they assess and in what proportion to fulfil specific human rights; they just have to be sufficient and diversified enough to protect individuals. This evaluation is undertaken by the Committee on ESCR, which has, for example, compared some states' military and health expenses to measure their priorities and resulting compliance with the right to health.[146]

If and when the new draft additional protocol is adopted, states will have to comply with a more precise inquiry procedure where some committees will assess compliance with their positive duties towards implementing the Covenant's right following complaints from groups or individuals.

Another issue with positive duties to implement socio-economic rights relates to the separation of powers and respective roles of the legislative, executive, and judiciary branches. Some view these as completely independent and consider that only the executive can initiate changes in the law to better comply with welfare human rights.[147] For these people, governments have a

---

144 A. Attaran, 'Human Rights and Biomedical Research Funding for the Developing World: Covering State Obligations Under the Right to Health' (1999) 4:1 *Health Hum Rights* 26.

145 R. Robertson, 'Measuring State Compliance with the Obligation to Devote the Maximum Available Resources to Realising Economic Social and Cultural Rights' (1994) 16 *Human Rights Quarterly* 693, at 703–713.

146 The Committee has done this in the past for Chile, UN doc. E/C.12/1988/SR.13, para. 12 and North Korean UN doc, E/C.12/1987/SR 22, pp. 5 and 17.

147 For example, see R. Calland and M. Taylor, 'Parliament and the Socio-Economic Imperative – What is the Role of the National Legislature?' (November 1997) 1 *Law, Democracy and Development* 193.

duty to identify priorities and take action to bring the object of the rights to the rights-holders; the judiciary only become involved afterwards, to enforce already-established mechanisms.[148] This vision appears too limited as it does not take into account the potential creative role the courts can play in the reinvention and protection of socio-economic rights. Indeed, judicial review can be seen as a crucial tool for ensuring the development of the normative content of socio-economic rights and their full and dynamic realisation.[149]

Negative duties Limited negative duties emerge in reaction to the vagueness of human rights and associated unspecified obligations. Pogge is one of the main advocates of the view that human rights give rise to negative duties not to harm others through the imposition of unfair institutional orders on them.[150] For Pogge, institutional orders should be evaluated according to the effect they have on the fulfilment of human rights. This is compatible with art. 28 of the UDHR and with Darwin's statement that, 'if the misery of our poor be caused not by laws of nature but by our own institutions, great is our sin'.[151] Pogge therefore supports an institutional conception of rights that sanctions claims against institutions only, as opposed to an *interactional* conception that would accept claims against anyone capable of satisfying the rights.[152]

As discussed previously under our assessment of the global order, current global institutional arrangements could be seen as a massive, collective infringement of human rights, especially socio-economic rights, by many of the world's most powerful and affluent agents.[153] Indeed, governments of the most developed countries and important multinational corporations are the

---

148 N. Haysom, 'Constitutionalism, Majoritarian Democracy and Socio-Economic Rights' (1992) 8 *SAJHR* 451, at 456.

149 S *v Makwarryane and another* 1995 (3) SA 391 (CC), 1995 (6) BCLR 665 (CC), at para. 325; IACHR Res. No. 12/85, Case 7615 (*Yanomami Indians v Brazil*), 5 March 1985, reprinted in Annual Report of the IACHR 1984–85, OEA/Ser.L/V/II.83, Doc. 14, corr. 1, at 33, 1 October 1985; *Eldridge v British Columbia (Attorney General)* [1997] 3 S.C.R. 624, 1997 CanLII 327 (S.C.C.); *Chaoulli v Québec (Attorney General)* [2005] 1 SRC 791.

150 Pogge deliberately refuses to take a stand on positive duties, even if he does say that he agrees with indirect positive duties to aid when people's basic human rights are at risk. T.W. Pogge, 'Severe Poverty as a Violation of Negative Duties', *supra* chapter 2, note 91, at 65–66.

151 Cited in S.J. Gould, 'The Moral State of Tahiti – and of Darwin' (1991) 10 *Natural History* 12, at 19.

152 T. Pogge, 'Cosmopolitanism and Sovereignty' (October 1992) 103 *Ethics* 49, at 50–51; T. Pogge, 'How Should Human Rights be Conceived?' (1995) *Jahrbuch fur Recht und Ethik* 3.

153 T. Pogge, 'Recognized and Violated by International Law', *supra* Introduction, note 3.

main artisans behind the functioning of the global order. This order, through numerous international treaties and agreements on trade, labour, intellectual property protection and investment, shapes most international economic transactions and contributes to the production of serious socio-economic inequalities and human rights violations.[154] For Pogge, the governments of affluent countries, and the citizens who elect and empower them, share a responsibility for human rights violations that unjustly disadvantage less affluent countries and their people, when these disadvantages are foreseeable and avoidable with practicable reforms. Instead of arguing that human rights encourage individualism and promote western values, the proponents of negative duties consider associated socio-economic human rights as individual moral claims on coercive institutions and on those involved in upholding them. Therefore, even if they do not have an individual positive duty to fulfil everyone else's basic rights, the emphasis on duties should encourage individuals to behave properly toward others and exert pressure on their representatives to respect their national and international duties.[155]

The negative duty not to impose and uphold an unfair institutional order can be seen as universal. It generates limited and definite positive obligations on the part of states and, sometimes, on citizens. States, empowered by citizens, must create effective institutions (or support and preserve existing ones), undertake reforms if those institutions are not adequate, and compensate those whose human rights are not fulfilled under the existing global order. Compared to sporadic and voluntary donations from wealthy countries, structural and institutional reforms would offer long-term consistency and fairer cost division among countries. Such reform should be undertaken to bring the global scheme to a level of justice where people could not be deprived of their right to health and where prevention and screening of serious medical conditions would be undertaken to meet a threshold of basic health needs – all in the broader framework of an ideal of equality of opportunities. To this end, respect for state sovereignty should be conditional on those states meeting minimal compliance with the protection of universal basic rights.[156] I will come back to the main obstacles to the efficient implementation of such negative duties at more length in the next section of this chapter, when I analyse the conceptualisation of human rights within the broader reality of the market.

This subsection has highlighted the imperfect nature of the duty and responsibility components of the human rights discourse. We have seen that, although positive and negative duties exist and can be allocated to different

---

154 Ibid.

155 H. Shue, *Basic Rights*, *supra* chapter 2, note 63, at. p. 131; C. Brown, *supra* Part 1, note 1.

156 H. Shue, *Basic Rights*, *supra* chapter 2, note 63, at 158 et seq.

agents, they are often difficult to comply with and to enforce on states through judicial processes. For example, a requirement to meet extraterritorial obligations related to human rights to health can infringe on state sovereignty (both on the giving and receiving end); achieving a balance between human rights protection and respect for state sovereignty can be somewhat challenging. This corresponds with the limited duty to fulfil (provide and promote) the legal right to health, that only requires states to undertake specific acts (reject discrimination and enforce minimum core obligations)[157] to realise the right in question, to the maximum of their available resources. However, as mentioned earlier, even when positive duties have not yet been allocated to specific persons or agencies, and even when rights are hard to realise, they can still exist and have great influence on how things evolve.[158] In other words, the imperfections of the duties discussed should not discredit the whole human rights discourse but are certainly symptomatic of a greater malaise.

In this last section, we discussed issues of access to health and distributive justice in connection with socio-economic human rights and related duties. We realised that although the system in place elevates health to the status of a universal value in international law and is supposedly designed and aimed at enforcing responsibilities in relation to this goal, things do not exactly work this way in practice. This is exemplified by the fact that the most deprived, health-wise, are neither protected nor taken in charge under this system; other, more powerful and affluent agents take control of the distribution of health and health-related goods and services, without having to acknowledge and respect the universal importance of health.

This clearly demonstrates the importance of taking our analysis one step further and undertaking a deeper evaluation of the complex global political and economic context within which the system of socio-economic human rights evolves. As Shue states, '[k]nowing how to protect the right against violation, or to restore the right after violation, depends as well on historical and empirical understanding of the relevant social, economic, political, legal, and psychological factors'.[159]

---

157 Maastricht Guidelines, *supra* note 12, guidelines 8, 9 and 10.

158 A. Sen, 'Consequential Evaluation and Practical Reason' (2000) 17:9 *The Journal of Philosophy* 478, at 495–498.

159 H. Shue, *Basic Rights*, *supra* chapter 2, note 63, at 158.

4.3 THE CONCEPTUALISATION OF HUMAN RIGHTS WITHIN THE REALITY OF THE MARKET

Rights can never be higher than the economic structure of society and its cultural development conditioned thereby.[160]

In this last section, I aim to present a more complete and balanced understanding of the system reviewed in the previous sections of this chapter. Outwardly, the human rights system seems to consider health as a universal value, asking for equal treatment of every human being in accordance with a cosmopolitan approach, and for the provision of an adequate amount of health care to protect equality of opportunities, regardless of socio-economic factors. However, in reality, fulfilment of positive and negative duties to secure a basic level of health for all has not occured, raising doubts about the true universality of human rights principles. There seems to be a gap between acknowledging such universal values in the form of human rights and giving them a real voice and impact within the economic and political reality of the world. I touched on some of those issues when I analysed the global order under which human rights develop, but here I go one step further and question the real nature of human rights – the interests lying at the basis of the system, both in terms of human rights content and implementation.

4.3.1 How is the Institution of Human Rights Shaped by the Market and the Powers in Place?

Different modes of social organisation exist, and the most important and influential in the current world order is the market. States which adopt an ideology that considers the market as the best way to distribute goods and services consent to limit their intervention and to prioritise privatisation and economic development.[161] For Evans, the ideology of the market refers to a set of normative relationships that exist without coercion, with a global reach, supported by discourses of truth, and widely accepted as 'common sense'.[162] Within this conceptualisation, rights like liberty, property, and free markets, which best contribute to secure important production and exchange, are often

---

[160] K. Marx and F. Engels, *Selected Works Vol. Three, 1875–1895* (Moscow: Foreign Languages Publishing House, 1970) at p. 19.

[161] S. Gill, 'Market Civilisation and Disciplinary Neoliberalism' (1995) 24:3 *Millennium* 412.

[162] T. Evans, *supra* note 30.

preferred to other more demanding and less economically-rewarding welfare rights.[163]

Different theoretical and practical arguments in favour of free markets exist and have been put forward in the literature. Those who endorse such a vision of the world typically maintain that minimally regulated international markets remain the best instruments for fostering innovation, technological development, individual freedom, democracy and optimal distribution of resources worldwide.[164] A deep analysis of this position is beyond the scope of this book. However, the following sections demonstrate why one cannot rely on the market to attain global distributive justice in health and that, in fact, the market frequently leads to economic and health inequalities.

The role of globalisation The key barrier to the realisation of socio-economic rights like the right to health is related to persistent gross economic inequalities observed within and between nations.[165] This is exemplified by the fact that globalisation is often not managed in the interest of developing countries and their people. Indeed, although globalisation has been beneficial for some countries which have chosen to gradually liberalise trade, most developing countries do not choose their own terms of participation, but have to comply with what the most powerful dictate. Due to the lack of democratic supervision at the global level, and because of the type of market pursued by the most powerful agents, strong economic interests tend to be prioritised, and agents with different priorities tend to be excluded, with negative consequences for their long-term interests.[166] The neo-liberal ideology promoted by developed countries in their international negotiations demands that markets be driven by efficiency, which means that 'concerns about any resulting poverty or inequality are externalised from the debate over markets'.[167] In such a context, distributive justice in health is not a priority at all. Indeed, providing appropriate goods and services to meet a threshold of basic health

---

163 British Medical Association, *The Medical Profession and Human Rights: Handbook for a Changing Agenda* (London: Zed Books, 2001), at pp. 24–26.

164 For more on different schools of thoughts related to the market theory see: M. Friedman, *Capitalism and Freedom* (Chicago: University of Chicago Press, 1962); A. Smith, *Adam Smith's Wealth of Nations: a New and Condensed Edition* (New York: T.Y. Crowell, 1904); J.E. Stiglitz and L. Squire, 'International Development: Is It Possible?' (Spring 1998) 11 *Foreign Policy* 138.

165 A. Eide, *supra* note 3, at 555.

166 A good example is taken from the Uruguay Round negotiations. For a more complete discussion on this, see J.E. Stiglitz and A. Charlton, *Fair Trade for All, How Trade can Promote Development* (New York: Oxford University Press, 2005).

167 K. Rittich, 'Transformed Pursuits: The Quest for Equality in Globalized Markets' (Spring 2002) 13 *Harv. Hum. Rts. J.* 231, at 257.

ends up being dealt with as a matter of sporadic assistance or charity, constrained by other requests and available resources. Socio-economic rights are therefore prevented from realising social welfare improvement, and can easily become empty provisions without effect.[168]

Some advocates of globalisation believe that human rights can only be realised through mechanisms of globalisation, as they both imply the same common language and associations among individuals all over the world.[169] Although critical of some of the possible negative effects of globalisation, international bodies have refrained from condemning it altogether.[170] However, globalisation has been heavily criticised by many organised social groups and academics from various disciplines,[171] who have nevertheless not succeeded in challenging and eradicating the impact that these forces have on the realisation of human rights. Market efficiency measured through protection of private property, contract enforcement, and a stable investment environment remains a priority and is the main factor used to assess policy initiatives.[172] This is in part due to the power of the strong financial alliance Bhagwati refers to as the 'Wall Street Treasury Complex', which represents a conglomeration of international financial institutions (IFIs), the US Treasury and State Departments, and Wall Street which greatly influence globalisation with their strategic actions, confound their interests with the interests of the whole world, and consequently threaten the substance of human rights.[173] Baxi is clearly addressing this when she says:

> I believe that the paradigm of the Universal Declaration of Human Rights is being steadily supplanted by a trade-related, market-friendly, human rights paradigm. This new paradigm reverses the notion that universal human rights are designed for the dignity and well being of human beings and insists, instead, upon the promotion and protection of the collective rights of global capital in ways that 'justify' corporate well-being and dignity over that of human persons.[174]

---

168 Ibid.

169 M.A. Warner, 'Globalization and Human Rights: An Economic Model' (1999) 25 *Brook. J. Int'l L.* 99.

170 A good example of this is art. 14 of the 1995 *Copenhagen Declaration on Social Development*, which notes some worries about globalisation, but at the same time also mentions its potential positive effects on economic growth and progress in developing countries.

171 R. Wade, 'Japan, the World Bank and the Art of Paradigm Maintenance: The East Asian Miracle in Political Perspective' (1996) 217 *New Left Rev.* 3.

172 K. Rittich, 'The Future of Law and Development' *supra* note 91.

173 J. Bhagwati, 'The Capital Myth: The Difference Between Trade in Widgets and Dollars' (May–June 1998) 77:3 *Foreign Aff.* 7, at 10–12.

174 U. Baxi, 'Voices of Suffering and the Future of Human Rights' (1998) 8 *Transnat'l L. & Contemp. Probs.* 125, at 163; cited in A. Anghie, *supra* note 93, at 249.

Powerful market actors and their impact on human rights As mentioned earlier, as things stand now, IHR implementation is almost exclusively national due to the relatively low priority that powerful countries award to other nations' human rights issues in their foreign policy agendas.[175] This being said, even when they make domestic moves in relation to human rights, states often find themselves driven by a larger agenda: helping and supporting the global economy built on the market ideology.[176] This means that the real importance given to the universal values enshrined in IHR treaties mainly depends on their compatibility with the overall purposes of the market. In other words, human rights end up being defined by powerful agents who often argue for a narrow conception, often only including civil and political human rights.[177] This way, the fact that socio-economic human rights violations are often caused by powerful market forces is not addressed in the dominant legal human rights discourse; this failure takes our attention away from the universal values enshrined in legal human rights in an insidious way.[178] This can bring people to wrongly believe in a just world where human rights are valued and respected when, in reality, so many suffer from serious deprivation at so many levels.[179] This could change when and if the draft optional protocol is adopted as groups and individuals claiming to be victims of a violation of their covenant's rights will have mechanisms to oppose their government and voice their human rights-related concerns and priorities.[180]

Another aspect of the contemporary world order is the tremendous power of transnational corporations, which now affects every state's ability to control its socio-economic agenda, even within its own borders.[181] States face strong pressure to adopt efficiency as their top priority and this ends up affecting different sectors of their activity, such as labour and trade. As Kothari puts it,

---

175 C. Brown, *supra* note 27.

176 R.W. Cox, 'Civil Society at the Turn of the Millennium: Prospects for an Alternative World Order' (1999) 25 *Rev. Int'l Stud.* 3.

177 B.S. Chimni, *supra* note 89, at 10.

178 T. Evans, *supra* note 30.

179 M.J. Lerner, *The Belief in a Just World: A Fundamental Delusion* (New York: Plenum Press, 1980).

180 Indeed, by signing and ratifying this protocol, member states would recognise the competence of the Committee on Social Cultural and Economic Rights to receive, consider and analyse communications related to human rights violations presented by groups and individuals. For more details on the protocol's content and functioning refer to: Revised Draft Optional Protocol to the International Covenant on Economic, Social and Cultural Rights, (24 December 2007) A/HRC/8/WG.4/2.

181 T. Friedman, *The Lexus and the Olive Tree* (New York: Farrar Straus Giroux, 1999).

'[c]apitalism is entering a new phase and economic processes are becoming autonomous of political authority'.[182] There is also a growing presence of private corporations within the UN structure. The *Global Compact*, an initiative of Kofi Annan that encourages responsible corporate actors to get involved in finding solutions to the challenges of globalisation, illustrates this.[183] Moreover, increasing corporate contribution to UN financing is reflected in the management philosophy adopted by the organisation and, consequently, 'reduces the possibility of UN forums being at the center of collective action by third world states to constrain these giant private actors'.[184] Since transnational corporations can exert a powerful influence on the socio-economic framework of states and of the global order, they can have positive and negative effects on the realisation of human rights.[185] However, most powerful transnational corporations have consistently refused to take any responsibility for the negative effect they may have on human rights. This is why some argue that they should, like states, be held accountable for human rights abuses not only through voluntary codes of conduct, but also through national and international regulations.[186] Some have proposed to establish, through international consensus, a governing body to act as a kind of international court to examine corporate actions.[187] However, many are sceptical about the practicability of such a project, at least as long as real power remains in the hands of a few influential corporations. They suggest focusing instead on civil actions and media exposure to encourage public stigmatisation of private economic actors, when required.[188] Indeed, the expanding social movement can make power visible while playing an important role in questioning power structures in a form of

---

182 R. Kothari, 'Globalization: A World Adrift' (1997) 22 *Alternatives* 227, at 228.

183 To learn more about the Global Compact, its functioning and its progress, refer to this website: http://www.unglobalcompact.org/ (accessed 2 June 2006).

184 B.S. Chimni, *supra* note 89, at 15.

185 S.R. Ratner, 'Corporations and Human Rights: A Theory of Legal Responsibility' (2001) 111 *Yale L. J.* 461; on potential positive effects of TNCs on developing countries refer to W.H. Meyer, *Human Rights and International Political Economy in Third World Nations: Multinational Corporations, Foreign Aid, and Repression* (Westport: Praeger, 1998).

186 G. Meintjes, 'An International Human Rights Perspective on Corporate Codes' in O.F. Williams (ed), *Global Codes of Conduct: An Idea Whose Time has Come* (Indiana: University of Notre Dame Press, 2000) 83; S.R. Ratner, 'Corporations and Human Rights: A Theory of Legal Responsibility' (2001) 111 *Yale L. J.* 461.

187 K.T. Jackson, 'A Cosmopolitan Court for Transnational Corporate Wrongdoing: Why its Time has Come' (May 1998) *J. Bus Ethics* 758.

188 M. Winston, 'NGO Strategies for Promoting Corporate Social Responsibility' (2002) 16 *Ethics & Int'l Aff.* 71.

'globalisation from below'.[189] As Stammer notes, '[t]here is a possibility that under contemporary conditions of globalisation – social movements might become more effective agents of global socio-cultural change in respect of human rights than existing nation-states and emerging supranational institutional structures'.[190]

The role of economic development for the protection of socio-economic rights

The most affluent and powerful agents (countries, private corporations, and international economic organisations) often argue that they contribute to protecting and fostering socio-economic rights in developing countries through their development and economic growth initiatives. However, this is not always true. In fact, most loans from international financial institutions are targeted to specific projects often unrelated to basic subsistence needs, health care, and education; when they are, these loans are part of 'adjustment lending processes', which frequently target decentralising and privatising reform initiatives.[191] These projects have greatly reduced developing nations' capacity to establish social programs compatible with their level of development.[192] Moreover, in focusing on their main creditors' short-term demands and interests, international financial institutions do not pay enough attention to the importance and role of investment in meeting basic health needs to improving many other sectors of economic activity, like employment.[193] Therefore, international financial institutions often end up acting like charitable lending organisations which have great powers of reform but do not necessarily use

---

189 To learn more about the practical and political role and influence of social movements, refer to André C. Côté, *Comment envisager la démocratie sans les groupes d'intérêts?* (May 2007) Université Laval, Colloque de la Société Québécoise de Science Politiques, online: http://www.commissairelobby.qc.ca/documents/File/ allocution_2007-05-24.pdf (18 March 2008); Gabriel A Almont, 'Research Note: A Comparative Study of Interest Groups and the Political Process' (1958) 51:2 *American Political Science Review* 270; Michel Offerlé, *Sociologie des groupes d'intérêts*, 2nd edition, Paris: Montchrétiens (1998).

190 N. Stammer, *supra* note 25.

191 Indeed, many argue that structural adjustment programmes established by international financial institutions have had negative effects on states' abilities to meet their human rights obligations. For more on this argument, refer to: M. Chossudovsky, *The Globalisation of Poverty: Impacts of IMF and World Bank Reforms* (London: Zed Books, 1997); M. Chossudovsky, *The Globalization of Poverty and the New World Order*, 2nd edn (Shanty Bay: Global Outlook, 2003).

192 J. Braithwaite and P. Drahos, *Global Business Regulation* (Cambridge: Cambridge University Press, 2000).

193 Indeed, promoting employment is one of the IMF's priorities. Articles of Agreement of the International Monetary Fund, 27 December 1945, art. 1, para. 3, 2 U.N.T.S. 39 on the importance of maintaining high levels of employment.

them to invest in nations' sustainable growth. They can consequently fail to acknowledge (and thereby violate) some of the most basic human rights of the citizens of borrowing nations.[194] This engenders and amplifies the horrible situation prevailing in many developing countries where children and young adults die every day of preventable and curable diseases partially associated with a tremendous debt repayment burden and a critical lack of public spending on health.[195]

Also, if we talk about private investors, given their mode of operation and their need for quick results, their notion of 'economic growth' will not necessarily result in needed long-term infrastructure investment, the promotion of employment and worker safety, initiatives that ensure environmental protection, or investments that truly help the citizenry and contribute in building their economy.[196] The initiatives supported by these actors, even if they objectively improve a state's circumstances with certain economic measures, will most often increase inequities and poverty among the vulnerable.

This clearly demonstrates that economic growth fostered by powerful national, international, and transnational agents, and often only targeted at improving macroeconomic variables, cannot automatically be associated with the realisation of socio-economic rights like the right to health.[197] Economic growth is too rarely evaluated in terms of capability and enhancement of human choices.[198] As a result of intense pressure from global civil society, international financial institutions have begun to approach these issues differently in promoting the social dimension of development for its positive effects on economic growth.[199] However, even if welfare goals are acknowledged,

---

194 This problem is best illustrated by what has happened in Latin America in recent years. For more on this, refer to J. E. Stiglitz, *Globalisation and its Discontents, supra* chapter 1, note 83; see also more generally on ESCR violations: J. Oloka-Onyango, 'Beyond the Rhetoric: Reinvigorating the Struggle for Economic and Social Rights in Africa' (1995) 26 *Cal. W. Int'l L.J.* 1, at 20–26.

195 Indeed, in Africa, many countries have to spend four times more money on debt repayment than they do on education and health care for their people. For more discussion and statistics on this, refer to A. Anghie, *supra* note 93, at 257–258; D. Ransom, 'The Dictatorship of Debt' (October 1999) *World Press Rev* 6.

196 UNDP, *Human Development Report 1995*; R. McCorquodale and R. Fairbrother, *supra* note 69.

197 In that sense, the *Human Development Index* is a much more complete indication of real growth as it highlights differences between economic growth and welfare.

198 M. Monshipouri, C.E. Welch and E.T. Kennedy, *supra* note 79, at 967.

199 For example, see World Bank, *International Bank for Reconstruction and Development, Development and Human Rights: The Role of the World Bank*, Washington, 1998, at 2, online on the World Bank website: http://www.worldbank.org/html/extdr/rights/hrtext.pdf (accessed 4 March 2009); see also IMF, *Guidelines on*

their reach is considerably restrained as they are defined within 'market promoting parameters' and ranked accordingly; equity and justice are important as long as they contribute to economic development. Liberalisation and privatisation remain the main tools for *securing* socio-economic rights.[200] The ICESCR notes, however, that those who signed and ratified the treaty have a human rights obligation of international assistance. Those states should therefore all work toward full implementation of the Covenant, even when they act through international bodies like the IMF or the World Bank.[201] Although every State party to the treaty has this obligation, it can be challenging for smaller and less affluent countries to get their ideas across in the process of implementing economic, social, and cultural rights, as international financial institutions use a voting scheme weighted by economic contribution, considerably advantaging the affluent northern nations.

In the current political world order, legal human rights end up directly supporting powers in place[202] while the UN represents the main actor engaged in the promotion of this neo-liberal agenda on the global scene. Although all human rights are meant to be equal and interrelated, in reality, respecting civil and political rights will be given absolute priority and socio-economic issues will only be addressed later, if they are addressed at all.[203] Alston notes that,

> In the world of globalization, a strong reaction against … the denial of primary education or health care, can often require not only showing that the relevant practices run counter to human rights standards but also a demonstration that they are offensive to the imperatives of economic efficiency and the functioning of the free market … [I]n order to be validated, a purported human right must justify its contribution to a broader, market-based 'vision' of a good society.[204]

This last section has demonstrated that although IHR law theoretically aims precisely to address and eliminate socio-economic inequities, those same inequities are caused by more dominant forces of globalisation. In this system, the promotion of the market is the absolute goal even if it does not further welfare and justice as powerful agents prefer to stay in their advantageous position and not lose some of their power. Also, as I have noted, the faults and

---

Conditionality, 25 September 2002, online on the website of IMF: http://www.imf.org/External/np/pdr/cond/2002/eng/guid/092302.htm (accessed 4 March 2009).

[200] F.J. Garcia, 'The Universal Declaration of Human Rights at 50 and the Challenge of Global Markets: Trading Away the Human Rights Principle' (1999) 25 *Brook. J. Int'l L.* 51; A. Anghie, *supra* note 93, at 262.

[201] S.S. Akermark, 'International Development Finance Institutions: the World Bank and the International Monetary Fund Agencies' in A. Eide et al. (eds), *Economic, Social and Cultural Rights: A Textbook*, 2nd edn (London: Maartinus Nijhoff).

[202] J. Donnelly, *supra* note 2, at p. 29.

[203] M. Mutua, *supra* note 70, at p. 35.

[204] P. Alston, *supra* note 88, at 442.

weaknesses of the human rights system in terms of structure and functioning are not disconnected from the fact that the system has been put into place by those same forces, in this same neoliberal context. It is therefore safe to say that socio-economic rights enforcement has been directly influenced by the commercial and political agenda of the most powerful agents of the world.[205] As powerfully summarised by Doyal and Gough: '[i]n assuming the state to be the key actor safeguarding human rights, the West's approach ignores the very real inequalities between states stemming from the political/military domination of the big powers and the economic dominance of the central capitalist states, financial institutions and corporations within the world economic order'.[206]

CONCLUSION

In this chapter, I analysed the IHR system to establish whether its underlying discourses, scope, structure, and functioning adequately account for the values encountered by my global distributive justice framework. My principal goal was to assess the human rights system to determine if it can be useful in the redistribution of potential genetic research benefits, taking health needs into consideration.

I realised that both discourses had strengths and weaknesses and identified another important facet of the human rights discourse: the political. Throughout the chapter, I evaluated the international human rights system with the benchmarks developed in my global distributive justice framework. I realised that the universalistic character of human rights could sometimes be used as a way to keep real people's needs at a distance, and that it could be criticised for its strong western influences. My assessment of the global context under which human rights evolve highlighted the strong capitalist roots that have led to a tolerance for economic disparity and injustice in health. I realised that the IHR system, through its institutional structure and its politics, does not necessarily protect the cosmopolitan approach to the solidarity and true universalism needed to support global distributive justice and access initiatives related to health and genetics.

Then, I studied how the human rights system deals with the crucial need for health and genetic-research access in terms of rights, entitlements, and related duties. My goal was to envision socio-economic rights with my benchmarks of distributive justice. I specifically analysed the human right to health in terms of validity, content, and efficacy; this led me to discuss associated

[205] K.E. Smith and M.M. Light, *Ethics and Foreign Policy* (Cambridge: Cambridge University Press, 2001).

[206] L. Doyal and I. Gough, *supra* chapter 1, note 42.

responsibilities to undertake distribution to facilitate access to health and genetic innovation. I realised that the concept of a 'right' was not the most useful in securing real access to health for the most needy and deprived, and that associated duties were hard to comply with and to enforce.

Finally, I was brought to realise that the real problem is with the powerful interests lying at the core of existing political and economic institutions that undermine the realisation of important socio-economic rights.[207] Legal norms therefore create 'entitlements' without considering the broader political and economic contexts created by the society's structures and institutions and which greatly influence decisions on resource allocation.[208] We should therefore work to alter these well-established economic forces if we are to rebuild trust in human rights, and create just distributive arrangements in the field of health and genetic technology. As Fidler states, '[j]ust as capitalism has become a truly global dynamic, the protection and promotion of health must also rise to the challenge of the new global order'.[209]

There are considerable political challenges to any reliable and sustainable implementation of socio-economic rights like the right to health. Even if socio-economic rights are legally protected, the whole context in which this happens helps safeguard capitalist market values, encourages passivity among the less affluent, and consequently discourages redistribution in health and other crucial sectors.[210] This highlights the enormous challenge of realising universal values of socio-economic equality and justice in a structure driven by a completely different agenda. This contributes to the lack of attention awarded to the ethical basis of socio-economic rights and to the false interdependence of human rights, all of which considerably weakens the practical implementation of the human right to health, as discussed earlier. In this sense, we have to agree with Kennedy when he says that the human rights movement can sometimes legitimate more injustice than it eliminates.[211]

This is what makes the system, as it currently operates, incompatible with my global distributive justice framework. If the human rights system were to meet my goal of global distributive justice in terms of global health, it would need to adopt an entirely different mode of functioning. This would require

---

207 K. Hossain, 'Globalisation and Human Rights: Clash of Universal Aspirations and Special Interests' in E. Burns, H. Weston and S.P. Marks (eds), *The Future of International Human Rights* (Transnational Publishers: New York, 1999).

208 K. Rittich, *supra* note 167, at 260.

209 D.P. Fidler, 'International Law and Global Public Health' (1999) 48 *Kansas Law Review* 1.

210 M. Mandel, *The Charter of Rights & the Legalization of Politics in Canada* (Toronto: Thompson Educational Publishing, 1994).

211 D. Kennedy, *supra* note 23, at 134.

undertaking institutional reforms to realise socio-economic human rights and, more specifically, to allow genetic technology and health-related goods and services to reach those for whom they can do the most good. To this end, '[w]e should not just assume that past forms of power will stay the same and have the same implications, nor should we assume that new forms of power will not arise'.[212] A fitting example of this is the growing social movement that has played an important role in generating pressure, resistance, and change in support of fairer resource distribution in various areas, sometimes in collaboration with international institutions and states.[213] However, to obtain effective and lasting results, they will need considerably more economic resources and much more political power to structure and rally a majority of people to their cause, particularly in the developing world.[214]

As more and more people come to realise that addressing the more pressing global health issues is critical for the creation of a more just and stable world order, developing efficient and enforceable human rights mechanisms appears to be one of the many important steps in the right direction. The draft optional protocol could be one instrument to help achieve this goal if it is adopted and efficiently enforced. However, until there is a major change in human rights politics, we will not be able to conclude that the system works to advance global distributive justice principles in terms of access to common health standards. This clearly highlights the important challenge associated with bridging the gap between the expression of universal values of cooperation, solidarity, and justice and the contingencies of modern world politics and economics.[215]

At the very beginning of this book, I presented evidence that benefits arising from genetic science have real potential for improving global health. In the first part, I argued that justice demands a broad redistribution of the benefits of genetics. I defended the equal and universal consideration of every individual's basic health needs to further a broader ideal of equal opportunity for all on the global scene. The analysis of the last two chapters reveals that the two main normative international systems meant to govern global access to scientific innovation and knowledge as well as distribution in health and genetic technology do not operate to advance equitable distributive justice ideals in

---

212 N. Stammers, *supra* note 25.

213 One example of this is the social ecologist movement, which has been very aggressive and efficient in challenging power relations and structure for a better and increased protection of environment rights. For more on this example, refer to B. Rajagopal, 'From Resistance to Renewal: The Third World, Social Movements, and the Expansion of International Institutions' (2000) 41 *Harv. Int'l L.J.* 529, at 533 and 578.

214 B.S. Chimni, *supra* note 89.

215 C. Brown, *supra* Part I, note 1; A.J. Langlois, 'Human Rights: the Globalisation and Fragmentation of Moral Discourse', *supra* note 47.

global health. My analysis has brought to light the magnitude of the gap between the normative expression of universal ideals of justice, equality, and solidarity, and the real limits imposed by the global world order's politics and economics.

As they currently function, the IP and human rights systems, taken both alone and together, do not allow adequate consideration of human welfare concerns. Despite the numerous positive law mechanisms in place, the philosophy and politics underlying these legal schemes hinder genetic-benefit redistribution and therefore prevent the realisation of global distributive justice ideals to broaden access to common health standards. This, however, does not mean that these normative systems should be completely abolished or replaced. They have definite potential, both individually and in concert, to contribute to global health improvement and allow individuals to profit from available opportunities. However, significant changes will have to occur in order to find some balance in applying these normative standards.

Establishing detailed solutions and practical policy options is beyond the scope of this book. In conclusion, however, I will say a few words about avenues that could be explored further, to set the basis for further discussion.

Conclusion

The primacy of human rights over trade liberalization is consistent with the trade regime on its own terms. The institutions that are the official guardians of trade law pose formidable barriers to the proper and full realization of this insight.[1]

As significant progress is being made in the field of human genetics, physicians, researchers and governments increasingly recognise that genetic technology, research tools, and therapeutic and preventive services are crucial for the improvement of global health. In particular, there is significant evidence that genetics will play an increasing role in medicine and public health in the coming years, and that it could consequently also have far-reaching impact on the health of developing countries' populations. However, less affluent countries often do not have the financial, technological and human resources to take advantage of these potential benefits and tailor them to their specific health care needs. Even with an increasing, globally accessible, body of scientific and technological knowledge and constant medical progress and discoveries, the condition of human health in many developing countries continues to decline.[2] Disparity in access to products and services arising from genetics is an important issue for the contemporary international policy agenda, and a specific challenge is to find ways to harness genetic knowledge so that it can contribute to global health equity through collaborative efforts. This topic has recently attracted a great deal of attention in many fora, especially in light of the widely-used concept of genetic-benefit sharing. Beyond outrage and intuitive feelings of injustice, however, the debate surrounding the global health and genetics divide needs to be brought one step further, through a deep analysis of the theoretical, legal, normative, socio-economic, and political factors involved in global inequalities.

---

1 R. Howse and M. Mutua, *Protecting Human Rights in a Global Economy: Challenges for the World Trade Organization*, Montréal, 2000, online on the website of Law and Democracy, http://www.ichrdd.ca/english/commdoc/publications/globalization/wtoRightsGlob.html (accessed 28 May 2009).

2 C. Juma and L. Yee-Cheong, 'Reinventing Global Health: the Role of Science, Technology and Innovation' (19 March 2005) 365 *The Lancet* 1105.

This book contributes to the debate about what has been called the genetics divide. In the face of the existing powerful market-oriented distribution mechanisms and the conceptual and normative weaknesses of the notion of compensatory benefit sharing, I adopted a different lens through which to analyse and reinvent the concept in relation to global health equity. To this end, I based my reasoning on the idea that justice demands the protection of the most vulnerable individuals to ensure that they benefit from equality of opportunities, an essential element for achieving justice.

The theoretical framework I have built covers many different aspects. I first established that an acceptable conception of justice in health should necessarily transcend boundaries. Indeed, while groups and communities are certainly very important units of consideration in many spheres of activity, I determined that we should deal with basic health and genetic needs using individuals as the ultimate unit of moral concern. This discussion resulted in the adoption of a cosmopolitan framework based on a global scheme of cooperation as the basis of my analysis. This global lens established a reference for assessing institutions that could be involved in the distribution of genetic benefits.

Following the establishment of my parameters of reference, I worked on a thorough analysis of the specificity of health to justify the elaboration of a particular framework of distributive justice in this area. Building on the work of Daniels and Rawls, this exercise highlighted the crucial importance of health and genetics for normal functioning, and the need to ensure a fair distribution of goods and services in this field because of the role normal functioning plays in individuals' ability to profit from available opportunities. This framework provided me with a basis for equal consideration of every person's health, for requiring equality of opportunities and, consequently, for the use of distributive justice schemes to solve avoidable health inequalities and encourage equitable access to the benefits arising from genetics rather than continuing to rely on market-based distributive mechanisms.

Following this theoretical analysis, I moved to the second goal of this book: assessing the compatibility of two important international legal systems concerned with distribution issues – intellectual property law and human rights law – with my framing principles. My investigation of the underlying philosophy, principles, structure and operation of these two legal frameworks led me to several conclusions. First, I realised that, although the intellectual property and the human right systems incorporate positive law dispositions relating to human welfare, knowledge diffusion and access, equity, and justice, neither system operates to advance equitable distributive justice ideals in terms of global health. I was faced with the conclusion that both structures, though quite different in their underlying rationales, are nevertheless similar in that they are both driven by powerful interests and market considerations in what we know as the global economy.

My work offers a significant contribution to the evaluation and analysis of this situation, beginning with the widely-used notion of benefit sharing, and questioning and restructuring it in such a way as to address and prevent a genetics divide. Doing so, I have produced strong and original normative landmarks that can be used to justify relying on a cosmopolitan approach to global justice based on health needs and opportunities in the face of major social, political, economic and legal pressure resulting from commercialisation of the fast growing field of genetics. My overall analysis resulted, however, in the identification of a clear gap between the framework I had constructed to give theoretical relevance to global benefit sharing obligations and the legal and political constraints that the application of two major and influential legal systems impose on global benefit sharing. For each legal system, I faced two kinds of deficit. First, I identified inherent legal (procedural and structural) problems that can impair the realisation of equitable distribution of genetic benefits. Secondly, I confronted broad, subtle, and major socio-economic and political problems affecting the functioning of the global order.

These findings represent some of the groundwork needed to initiate policy discussions and to eventually undertake concrete changes to achieve an international redistribution of resources emerging from genetics, and likely for other promising technologies with potential for global health improvement. With this grounding, we can begin to consider whether there exist any short and middle-term policy solutions that we could implement within the actual global normative architecture. More generally, we can start to think about how to approach the greater challenge of limiting hegemonic forces and powers that lie at the basis of this same global architecture to allow global distributive justice in health. Although establishing detailed solutions and practical policy options is not the aim of this work, by way of conclusion, I will, however, say a few words about tensions that would need to be resolved and avenues that could be explored further in the short and longer term, to lay some foundation for further discussion.

However, before doing so I have to acknowledge that many other non-legal, country-specific elements can also pose significant hurdles to global health improvement. Some important factors unrelated to patent and human rights that can influence genetic research and innovation in developing countries are their critical lack of research facilities, infrastructure, and expertise and their political instability.[3] It is often nearly impossible for most of them to use, distribute, and administer existing research instruments and technology,

---

[3] L. Bernier, K. Durell and E.R. Gold, *The Impact of DNA Patents on Access to Genetic Technologies and Services: View from Developing Countries*, World Health Organization, Geneva, January 2004, at p. 31 et seq; Groupe international d'experts en biotechnologie, innovation et propriété intellectuelle, *supra* chapter 3, note 70.

never mind genetic products and services that can be licensed or purchased from developed-world patent holders.[4] Although such factors are not always completely independent from legal systems[5] they are often related to domestic political and financial instability, issues to be addressed both internally and with external financial and educational input; in some cases in combination with, and, in others, independently of, the normative system. Addressing those issues in details was beyond the scope of this book but one has to remember that any sustainable realistic solution to global health issues will need to take those crucial factors into account.[6]

POLICY OPTIONS TO EXPLORE WITHIN THE EXISTING GLOBAL STRUCTURE

Concerning the human rights system, alternative measures could be established within the existing legal structure to address the poor level of enforcement and justiciability of socio-economic rights. For example, one author has proposed a 'concerted and integrated approach'[7] under which socio-economic human rights could be indirectly enforced though the application of civil and political human rights, or be considered as a specific social or ethical dimension of those same civil and political rights, given the clear link existing between the two types of rights and the better justiciability of the latter. This strategy would only emphasise and give concrete expression to the already well-recognised principles of indivisibility and inter-relatedness of rights. Another way of envisioning socio-economic rights implementation has been to argue for their progressive realisation through a 'minimum threshold approach'. This approach would justify compelling governments to defend

---

4 J.M. Spectar, 'Patent Necessity: Intellectual Property Dilemmas in the Biotech Dominant Treatment Equity for Developing Countries' (2001) 24 *Hous. J. Int'l L* 227.

5 For example, as previously discussed, intellectual property rights may influence countries' access to crucial research tools for building scientific and technological capacity. On this point: C. Juma and K. Fang, 'Bridging the Genetic Divide' in M. Ruse and D. Castle (eds), *Genetically Modified Foods: Debating Biotechnology* (Amherst: Prometheus Press, 2002).

6 For more on those factors and for some interesting recommendations to address some of those issues, refer to L. Bernier and K. Durell, 'Are Gene Patents Really a Threat?' in B.M. Knoppers and E.R. Gold (eds), *Biotechnology Intellectual Property and Ethics* (Montreal: Butterworth, forthcoming).

7 S.A. Agbacka, 'Reclaiming Humanity: Economic, Social and Cultural Rights as the Cornerstone of African Human Rights' (Annual 2002) 5 *Yale Human Rights and Development Law Journal* 177.

their priorities and require that they provide the minimum level of enjoyment of the whole range of basic human rights before prioritising any sector of economic activity.[8] Such an approach could be more effective with the establishment of specific indicators against which we could assess compliance with a minimum level of socio-economic rights.[9]

In terms of changing the patent system, patent pools have been proposed as a response to the potential negative effects of patents on fundamental research and access to technology. It has been suggested that the application of patent pools in the area of diagnostic genetics may work to provide greater access to genetics technology and to encourage collaboration between different agents involved in this sphere of activity who share the same goal of developing accurate, safe, and reliable testing methods for given polygenic diseases. However, industries may be reluctant to share their patents via a pool, preferring instead to pursue their research alone with the hope of bigger financial returns.[10] To solve this dilemma and encourage the formation of patent pools in genetics, inventive and attractive licensing and financial redistribution schemes could be set up to encourage industries to appreciate and acknowledge the financial and social advantages of patent pools.[11] UNITAID patent pool initiative will maybe offer some interesting ideas and srtategies once it is set up. Nevertheless, many other challenges remain to the feasibility of patent pools in genetics. More research and a more thorough analysis than can be offered here would be needed to assess whether, in the end, patent pools offer any real hope of addressing access concerns.

Innovative licensing strategies could also play an important role in securing access to genetics by developing countries.[12] Indeed, when licensing fees become prohibitive in terms of affordability of innovations, developing countries should think about other modes of securing access to patented technologies. As discussed briefly in the course of this book, one solution is to allow

---

[8] A. Bard-Anders et al., 'Assessing Human Rights Performance in Developing Countries: The Case for a Minimal Threshold Approach to Economic and Social Rights' in A. Bard-Anders and E. Asbjorn (eds), *Human Rights in Developing Countries* (Copenhagen: Akademisk Forlag, 1988) at p. 333.

[9] P. Hunt, *Interim Report of the Special Rapporteur of the Commission of Human Rights on the Right of Everyone to Enjoy the Highest Attainable Standard of Health*, 58th Sess., Agenda Item 117(c). A/58/427, 10 October 2003.

[10] A.K. Rai, *supra* chapter 3, note 30.

[11] T.J. Ebersole, M.C. Guthrie and J.A. Goldstein, 'Patent Pools as a Solution to the Licensing Problems of Diagnostic Genetics' (January 2005) 17:1 *Intellectual Property & Technology Law Journal* 6, at 11.

[12] L. Nelsen, 'The Role of University Technology Transfer Operations in Assuring Access to Medicines and Vaccines in Developing Countries' (January 2003) III:2 *Yale Journal of Health Policy Law and Ethics* 301.

compulsory licensing within national patent systems, as permitted by TRIPS under the heading not only of 'national emergency or other circumstances of extreme urgency' but also of 'public non-commercial use' (art. 31 TRIPS).[13] Some are of the view that these last-resort restrictions are very limiting and suggest that the scope of compulsory licences should be extended to help provide access to genetic tools and technologies for prevention purposes, for example.[14] I instead believe that the problem is not with the rights themselves but with how some countries are afraid to use and enforce them out of fear of retaliation from the most powerful countries. Another solution would be for governments to use mechanisms to encourage different industry sectors to agree on consistent and uniform advantageous licensing practices when dealing with developing nations. Such an agreement could create 'ethical business leadership' while helping countries to meet their international cooperation obligations and would likely not affect private profits substantially.[15] Apart from licensing, other mechanisms exist in IP law to facilitate access and are also widely debated and discussed on the international scene.[16]

In addition to these system-specific initiatives, more effort could be expended to better ensure compatibility and connection between human rights and intellectual property rights. The gap that makes them evolve on separate tracks could be bridged, at least partially, using different strategies, such as working towards a more organised and concerted action and strategy from different branches of the civil society; improving the dialogue between policy makers and governmental officials of the two sectors on the national and international scene; and focusing on the social role of IP rights, on their compatibility with human rights and on their related capacity to protect access to a certain level of health in priority. For example, in response to the major normative and political obstacles they

---

13 E.R. Gold and D. Lam, 'Balancing Trade in Patents: Public Non-Commercial Use and Compulsory Licensing' (2003) 6 *The Journal of World Intellectual Property* 5. The WTO's General Council adopted a decision on 30 August 2003 in Cancun, which allows countries without manufacturing capacity to also use compulsory licensing for importation rather than production.

14 C.M Correa, *Intellectual Property Rights and the Use of Compulsory Licenses: Options for Developing Countries,* Working Paper for the Center for Advanced Studies at the University of Buenos Aires, Argentina.

15 A. Attaran and L. Gillespie-White, 'Do Patents for Antiretroviral Drugs Constrain Access to AIDS Treatment in Africa?' (2001) 286:15 JAMA 1886.

16 I am referring, for example, to differential pricing and price control depending on the market and the consumers' ability to pay, imposing local working of patented products following their introduction in a given market, allowing R&D of generic products before the patent expires with Bolar provision, and allowing parallel imports of patented products. For an extensive discussion on those mechanisms, refer to J.H. Barton, *Differentiated Pricing of Patented Products, Commission on Macroeconomics and Health Working Paper Series*, Paper No. WG4:2, July 2001.

face with TRIPS' application, some developing countries, assisted by international NGOs and intergovernmental organisations, are now trying to further their interests in other fora, adopting strategies of regime-shifting. These initiatives aim to expand IP lawmaking in other regimes like biodiversity, public health, and human rights to address the social aim of IP and challenge and revise some of TRIPS' problematic dispositions and associated practices.[17] This can result in conflicting and contradictory legal obligations, and the results will depend on the nature of the emerging documents, binding or non–binding; their enforcement mechanisms; and the authority, mandate, and resources of the organisation in charge of implementing them.

As well as legal mechanisms, new types of business strategies and partnerships could also be explored and adopted to complement the human rights and the intellectual property rights systems. For example, in reaction to global health inequalities induced by market forces, some have proposed reliance on new ways to finance R&D in genetics, especially for neglected conditions, involving key players through public–private partnerships (PPP).[18] As an alternative to PPP, some have instead proposed a mandatory global *need tax* mechanism, which would be applicable to the profits of private agents in certain high-profile spheres of activity and subsequently redirected toward the needs of developing countries.[19] Another option is a *Global Resources Dividend*, requiring agents who exploit natural resources to compensate those who do not have the opportunity to make use of and profit from the same resources.[20] Another possibility would be to establish an international regime overseeing the pursuit of research and development activities and the distribution of benefits in the area of human genetics.[21] Instead of being exclusively

---

17 L.R. Helfer, *supra* chapter 3, note 231.

18 Such a strategy aims to get funds from the private sector to further R&D for global health issues with the ultimate purpose of providing affordable and adapted goods to populations from the developing world. To this end, IP arrangements have to be negotiated creatively to ensure that private companies have enough commercial incentive to invest and that IP can also be used as a tool in the pursuit of social objectives. Such an initiative has taken place in many spheres, including malaria, tuberculosis, and AIDS research with the Global Fund to fight AIDS, Tuberculosis and Malaria. See Groupe international d'experts en biotechnologie, innovation et propriété intellectuelle, *supra* chapter 3, note 70 at p. 29. E. Ziemba, *Public–Private Partnerships of Projects Development: Financial, Scientific and Managerial Issues as Challenges for the Future*, CIPIH Research Report, Geneva, 2005, online on the WHO website: http://www.who.int/intellectualproperty/studies/Ziemba.pdf (accessed 21 May 2009).

19 D.G. Richards, *supra* Introduction, note 31, at p. 141.

20 T. Pogge, 'Severe Poverty as a Violation of Negative Duties', *supra* chapter 2, note 91; T.W. Pogge, 'A Global Resources Dividend', in D. Crocker and T. Linden (eds), *Ethics of Consumption* (Totowa, NJ: Roman and Littlefield, 1999) p. 501.

21 It could follow the distribution model established in the recent International

driven by IP law, this system could be a stand-alone mechanism and be viewed both as an alternative and a complement to the present regime.[22] An additional area where attention could be focused is the improvement of public, government, media, and health professional awareness of the medical potential of genetics, the functioning of the IP system, and important international negotiations on complex technological and normative issues. This could take the form of a global network of different agents involved in the field with a strong presence of people from developing countries and the civil society.[23]

This being said, even if some negative consequences of socio-economic and health inequalities could theoretically be addressed through human rights law and intellectual property law built-in flexibilities and complementary policy and business strategies, it has become clear that neither system is currently working very effectively in equitably distributing benefits. Socio-economic rights are often perceived as 'feel good window dressing' rather than real, enforceable norms,[24] and IP flexibilities are frequently of very limited practical utility. This is not a problem, however, with the systems per se, but is instead symptomatic of the larger, troubling economic and political global reality analysed in this book.

Indeed, the patent system is intended, in part, to provide incentives for innovation, disclosure of invention, and, ultimately, increase public knowledge. IP law is not static, and if we go back to its roots and emphasise its role

---

Treaty on Plant Genetic Resources for Food and Agriculture FAO, International Treaty on Plant Genetic Resources for Food and Agriculture, *supra* Introduction, note 36. To this end, an international organisation could be in charge of the regime, collecting royalty fees from the use of the human genetic pool and setting up redistribution mechanisms on the basis of the impact of inventions on global health, for example. For more on what those systems could look like, refer to: A. Motoc, *Specific Human Rights Issues, Universal Declaration on the Human Genome and Human Rights*, UN Economic and Social Council, Geneva, 2002, doc.E/CN.4/sub.2/2002/37, paras. 13–15.

22 For example, it could include terms consistent with section 67 of TRIPS, which provides a basis for cooperation between affluent and less affluent countries while concurrently ensuring transfer of a portion of patent royalty fees to the researchers involved.

23 It could be similar to the latest regional initiative of the New Partnership for Africa's Development (NEPAD) or the *Global Genomics Initiative* proposed by the Joint Centre for Bioethics recently. For more information on these, see NEPAD and African Institute for Capacity Development (AICAD), *Memorandum of Understanding to collaborate on Capacity Development and Poverty Reduction in Africa*, Kenya, 4 November 2005; and Joint Center for Bioethics, *Genomics and World Health*, Toronto, 2004, at 42–44.

24 M. Darrow and T. Amparo, 'Power, Capture and Conflict: A Call for Human Rights Accountability in Develoment Cooperation' (May 2005) 27:2 *Human Rights Quarterly* 471, at 479.

in greater knowledge diffusion and access, it can contribute to improving the social good. In fact, the basis of the global trade system was established as part of a broader objective of global peace and security following World War II. This is clearly evident from art. XX of the GATT on the primary importance of protecting public morals and human life, and from the Preamble of the *Agreement Establishing the WTO*, which states that the purpose of the system is not free trade at all costs, but also 'allowing for the optimal use of the world's resources in accordance with the objective of sustainable development'.[25] As suggested by May, the public access and social utility aspects of IP should therefore be emphasised and elevated to the rank of primary consideration, while relegating the private characteristics to the status of privilege.[26] Similarly, the human rights system aims to address, limit, and solve different types of inequalities with the ultimate ideal of attaining universal respect for human beings' freedom, dignity, and equality. It is therefore important to emphasise, again, that equity and access problems are not created by the systems of IP rights and human rights per se, but partly by how they are manipulated by external agents driven by powerful economic and political interests.

Policy changes and isolated business strategies like those just addressed could certainly tackle problems arising with the application of human rights and IP law and improve global access to health and genetic benefits temporarily. However, limits to access and obstacles to distributive justice in health are bigger than intellectual property and human rights law and policy. In fact, we have been faced, on many occasions in the course of this book, with the limits of analysing legal processes independently of other social, political, and economical factors. Even if clear legal dispositions or court orders aiming at improving individual health through better access exist, their practical and real impact often depends on broader political and economic factors and struggles which originate in the very construction and functioning of the global order under which normative systems evolve.

BROADER LONG-TERM AND INCLUSIVE OPTIONS

Actions outside of the system of rights, at the level of global governance and international architecture, will be needed if we are to expect any significant, inclusive, and sustainable solutions to limit hegemonic forces, to include the most vulnerable and change the way human rights and intellectual property principles are translated into reality. Institutional changes of many kinds can

---

25 *Marrakesh Agreement Establishing the World Trade Organization*, Geneva, 15 April 1994.

26 C. May, *supra* chapter 3, note 1.

be envisioned and new influential actors of the global political picture, like international non-governmental organisations and transnational corporations, need to be involved.

We will need to reflect on establishing innovative strategies to give more space and authority to developing nations in international institutions.[27] This is an important challenge, due to the tremendous financial control that some powerful states operate on the policy positions, priorities, and initiatives of those institutions. As Richards states, 'insofar then as nation-state governments themselves are dominated by global capitalist interests, the difficulties of creating and financing transnational regulatory agencies that are autonomous of these same interests are magnified'.[28] To this end, attention could therefore be focused on the role and capacity of NGOs to influence the behaviour of the most powerful agents of the world and to change their established dynamic. At many occasions, NGOs of various regions and different social sectors have had a positive social impact through public awareness initiatives and widely-publicised opposition to human rights violations and injustices committed by the world's most powerful stakeholders.[29] Over the last decade, they have become increasingly organised, articulate, and powerful. They have been the source of various highly-publicised initiatives and campaigns for better human rights protection, including the national and international battles for affordable HIV/AIDS drugs and well-orchestrated resistance to international institutions and big pharmaceutical companies' actions.[30] Their capacity to 'investigate, expose and shame'[31] is their main strength. Unfortunately, the concrete results of their actions are often limited, given their restricted resources to further their social and political agenda and to network with other similar groups to establish concerted actions and have a voice in the international dialogue.

---

[27] A. Kwa, 'Power Politics in the WTO, Focus on the Global South' (second edition), January 2003, online on the Focus on the Global South website, http://www.focusweb.org/publications/Books/power-politics-in-the-WTO.pdf (accessed 11 April 2009); B.S. Chimni, *supra* chapter 4, note 89; J.E. Stiglitz, *Globalization and its Discontents*, *supra* chapter 1, note 83, at p. 225.

[28] D.G. Richards, *supra* Introduction, note 31, chapter 6, at p. 147.

[29] For an interesting overview of some successful NGO actions and strategies, refer to R. Khan, 'The Anti-Globalization Protests: Side-Show of Global Governance, or Law-Making on the Streets?' (2001) 61 *Heidelberg Journal of International Law* 323.

[30] B. Gellman, *An Unequal Calculus of Life and Death; As Millions Perished in Pandemic, Firms Debated Access to Drugs; Players in the Debate Over Drug Availability and Pricing* (27 December 2000) *Washington Post*, at A1.

[31] K. Roth, 'Defending Economic, Social and Cultural Rights: Practical Issues Faced by an International Human Rights Organization' (2004) 26 *Human Rights Quarterly* 63, at 67.

Agents from civil society and leaders from developing world governmental agencies are already in place, trying to coordinate their work and initiatives against inequalities and in furtherance of justice in distribution and access. They need to be empowered if we want them to effectively contribute to breaking the current health inequality cycle.[32] To this end, coalitions and strategic alliances between countries and organisations with the same distributive justice vision need to be established, and important reforms have to be envisioned. This can build on the ethical vision and international social purposes slowly spreading as a result of globalisation, what has been referred to as the 'social dimensions of globalization'.[33] Such a strategy has had mitigated success on the international scene up to now because of widespread resistance from some affluent nations to renouncing any of their sovereignty in favour of a more equitable global negotiation process. In response to this international deficit, countries and organisations with similar problems and visions could aim to form strong alliances at the regional level, for example, as a way to become less isolated, stronger, and more coordinated in their opposition to the commodification of health and genetics. This could be a first step in the process of gradually reforming the hegemonic international order to make it more accountable in the long term, not just in the health and genetic sphere, but in many other areas where technological development is likely to give rise to socio-economic inequalities and exclusions due to powerful market demands.[34]

Also, in the face of the immense and growing economic power of transnational corporations in this same global picture, another avenue could be to increase corporate social responsibility. Some oppose additional responsibilities, arguing that the first duty of transnational corporations (TNCs) is to make profit for their shareholders' benefit, while respecting the law of the states in which they operate.[35]

However, the perception and the role of TNCs is gradually changing. Indeed, they are more and more regarded as social organisations with social duties towards their employees, the environment, and society at large. TNCs

---

32 F. Fukuyama, *End of History and the Last Man* (New York: Free Press, 1992); M. Mandelbaum, *The Ideas that Conquered the World* (New York: Public Affairs, 2002); F. Zakaria, *The Future of Freedom* (New York: Norton, 2003).

33 J.A. Ocampo, 'Rethinking the Development Agenda' (2002) 26:3 *Cambridge Journal of Economics* 393.

34 For more on what has been called 'open regionalism', refer to J.A. Ocampo, 'Globalization, Development and Democracy' (2005) 5:3 *Items and Issues*, online on the website of Yale Global, http://yaleglobal.yale.edu/about/pdfs/ocampo.pdf (accessed 2 May 2009).

35 P.T. Muchlinski, 'Human Rights and Multinationals: Is There a Problem?' (2001) 77:1 *International Affairs* 31.

themselves seem to be inclined toward some social role, as most of them are now voluntarily adopting corporate codes of ethical conduct. Given this changing reality, insisting on TNCs' accountability for human rights violations and persisting inequities could contribute to gradually changing the actual global architecture and its troubling focus on the needs of its more powerful and wealthy agents.[36] One strategy could be to insist on the market value of a positive public image and on the effect that respect for human rights could have on a company's economic value.

Again, however, such reforms have to arise from a real and shared conviction that global health gaps and human rights deficits should be eliminated in priority, even if it means direct and coordinated contributions, changes in the distribution of benefits, and opportunity costs for the most affluent. It has to gain support from governments, corporations, and the public at large. In the actual global order, resolving the inequitable power distribution arising in the social, economic, and political arenas appears very challenging, since the most affluent are almost always the most powerful politically.

The current patterns of inequality created by the tremendous influence of powerful states and TNCs on the global order are neither natural nor unavoidable, but are the result of political choices driven by powerful socio-economic concerns for which no one seems to be held accountable. As highlighted by Howse and Mutua, the United Nations, WTO, and international financial institutions are actually in place and represent the institutional foundation needed to arrive at some agreement on complex issues touching on human rights, economics, and trade. These organisations are not incompatible with one another and could work in a collaborative way instead of evolving independently and inconsistently, as they often do. Institutional evolution, more accountability for non-state actors, and a broader perspective on trade law's implications are needed if we expect to bridge the gap existing between human rights and economic institutions. Such reorganisation will likely only happen through a reallocation of political authority, a very complex and long-term global project. Some, inspired by the recent successful grassroots environmental movement, suggest starting with a re-evaluation and valorisation of the public realm to rationalise restrictions on property rights.[37] In any case, a

---

[36] One example of such a proposal is the UN Global Compact discussed previously, where corporations agree to promote human rights in their spheres of activity. For a comprehensive discussion on the social responsibility of TNCs, refer to J.E. Parkinson, *Corporate Power and Responsibility* (Oxford: Clarendon Press, 1993); J. Dine, *The Governance of Corporate Groups* (Cambridge: Cambridge University Press, 2000).

[37] J. Boyle, *The Public Domain: Enclosing the Commons of the Mind,* supra chapter 3, note 26; C. May, *supra* chapter 3, note 1 at 144.

sense of the common interest in health has to develop across different and disparate groups to move toward global distributive justice in health.

More research and analysis should be invested in exploring such a strategy further, especially on how alternative political forces could be used and manipulated differently to ensure some sort of social control over knowledge, products and services access and distribution. For example, it will be interesting to study how existing political and social powers both from the developed and the developing world could mobilise, build a consensus and join together in their action to influence the international distributive scheme. Also, strategies will be needed to translate the cosmopolitan notion of global citizenship and the widespread refusal of health inequalities in the global dialogue and make them strong enough to resist market liberalisation and national political forces. More generally, it will be essential to continue thinking about the most equitable and realistic way to envision a redistribution of authority, governance and related institutional reforms at the global level.

Bridging the genetics divide requires more than an injection of money into innovative research projects and products relevant for the needs of developing countries. In this book, my approach has been focused principally on the normative responses and deficit in approaching the genetics divide but I have quickly realised that a broader vision is required if we want to tackle the major issues of health inequalities. Achieving justice in health and more equitable distribution of the benefits emerging from genetics to further human life and health improvement is of public interest and is within our capacity. It requires collective action towards building functioning innovation systems in developing countries and addressing local deficits including education, scientific capacity building, infrastructure improvement and corruption eradication. Ultimately, it mostly requires questioning the established global order and working towards reducing and eliminating persisting social inequalities by using national and global political and economical forces in a different way.

As a global community with resources, knowledge and technology to reduce and even eliminate the majority of existing global health issues, we have a responsibility to act to prevent radical inequalities.

Bibliography

LEGISLATION

African Charter on Human and People's Rights, adopted 27 June 1981, OAU Doc. CAB/LEG/67/3 rev. 5, 21 I.L.M. 58 (1982), *entered into force* 21 October 1986 Art. 16.

Argentina, Law 23.798 of 1990.

Brazilian Patent Law 9.279 of 1997.

Canadian Patent Act, Canada, art. 55.2.(1).

Charter of the United Nations, 892 *U.N.T.S.* 119, art. 2(1). art. 2 and 23 ICESCR; art. 55 and 56.

China Patent Law 1982, amended in 1992.

Constitution of the Bolivarian Republic of Venezuela, Dec. 1999.

Constitution of the Republic of South Africa (adopted 8 May 1996, amended 11 October 1996), online: http://www.polity.org.za/govdocs/constitution/saconst.html (accessed 3 June 2006).

Indian Patents Act, 1970.

Uruguay Presidential Decree of 1997.

JURISPRUDENCE

Airey v Ireland, (1979), 32 Eur. Ct. H.R. (ser. A) *reprinted in* 2 Eur. H.R. Rep. 305 (1979).

Apotex Inc. v Wellcome Foundation Ltd., (2002) 77 S.C.C.

Baker v Canada (Minister of Citizenship and Immigration), [1999] 2 S.C.R. 817.

Bermudez et al v Ministerio de Sanidad y Asistencia Social, (1999) Supreme Court of Justice of Venezuela, Case No. 15.789, Decision No. 916.

Cadbury Schweppes Inc. v FBI Foods Ltd, [1999] 1 S.C.R. 142.

Causa No. 31, Viceconte, Mariela Cecilia c/Estado NacionalMinisterio de Salud y Acciôn Social, s/Amparo Ley 16.986. Câmara Nacional en lo Contencioso-Administrativo Federal, Sala IX 2 June 1998.

Chaoulli v Québec (Attorney General), [2005] 1 S.R.C. 791.

re Dane K. Fisher and Raghynath v Lalgudi, United States Court of Appeals

for the Federal Circuit, 04-1465 (Serial No. 09/619,643), 7 September 2005.
Diamond v Chakrabarry, (1980) 447 U.S. 303, 100 S. Ct. 2204.
Eldridge v British Columbia (Attorney General), [1997] 3 S.C.R. 624.
Embrex Inc. v *Service Engineering Corp.*, 216 F.3d 1343, 1349, 55 U.S.P.Q. 2d 1161, 1163 (Fed. Cir. 2000).
Fogerty v Fantasy Inc., 510 U.S. 517 (1994).
Graham v John Deere Co., 383 U.S. 1 (1966).
Greenberg v Miami Children's Hospital Research Institute Inc., 208 F. Supp. 2d 918 (N.D. Ill. 2002). 121. 793 P.2d 479 (Cal. 1990).
International Federation of Human Rights Leagues (FIDH) v France, (complaint no. 14/2003).
Madey v Duke University, 307 F3d 1351 (Federal Circuit Court of Appeal 2002).
Merck KGAA v Integra Lifesciences I, Inc., (2005) W.L. 1383624, *1, 8 (U.S. 13 June 2005).
Moore v Regents of the University of California, 794 P 2d 479 Cal SC 1990.
S. *v Makwarryane and another,* (1995) (6) BCLR 665 (CC).
Schmeiser v Monsanto Canada Inc, [2004] 1 S.C.R. 902.
Schneider v The Queen, [1982] S.C.R. 112.
Soobramoney v Minister of Health, Constitutional Court of South Africa, (1997) 12 BCLR 1696 (CC).
Tavares v France, App. No. 16593/90, Eur. Comm'n H.R. Dec. & Rep. (12 September 1991).
Treatment Action Campaign v Minister of Health, High Court of South Africa, Transvaal Provincial Div., (2002) (4) BCL 356M.
XXX v Ministry of Health, Supreme Court of Argentina, 1 June 2000.
Yanomami Indians v Brazil, Mar. 5, 1985, IACHR Res. No. 12/85, Case 7615 (reprinted in Annual Report of the IACHR 1984-85, OEA/Ser.L/V/II.83, Doc. 14, corr. 1, at 33, 1 Ocober 1985).

SECONDARY MATERIAL: MONOGRAPHS

Addo, M.K., *Human Rights Standards and the Responsibility of Transnational Corporations* (Boston: Kluwer Law International, 1999).
Andersen E.A. and B. Lindsnaes, *Towards New Global Strategies: Public Goods and Human Rights* (Netherlands: Martinus Nijhoff Publishers, 2007).
Anderson, J. et al., *Malaria Research: An Audit of International Activity* (London: The Wellcome Trust, 1996).
Andrews, L.B., *Future Perfect* (New York: Columbia University Press, 2001).

Angell, M., *La vérité sur les Compagnies Pharmaceutiques* (Montebello: Les Éditions le mieux-être, 2005).

Baker, R., *Capitalism's Achilles Heel: Dirty Money and How to Renew the Free-Market System* (New Jersey: John Wiley & Sons, Hoboken, 2005).

Ball, T. and R. Dagger, *Ideals and Ideologies: a Reader* (New York: Pearson Longman, 2004).

Barry, B., *The Liberal Theory of Justice: Critical Examination of the Principal Doctrines In A Theory of Justice by John Rawls* (Oxford: Clarendon Press, 1975).

Becker, L.C., *Property Rights: Philosophic Foundations* (London: Routledge, 1977).

Beitz, C.R., *Political Theory and International Relations* (Princeton: Princeton University Press, 1999).

Bentham, J., *Rights, Representation, and Reform: Nonsense upon Stilts and other Writings on the French Revolution* (New York: Oxford University Press, 2002).

Bowles, S. and H. Gintis, *Democracy and Capitalism: Property, Community, and the Contradictions of Modern Social Thought* (New York: Basic Books, 1986).

Boyle, J., *The Public Domain: Enclosing the Commons of the Mind* (New Haven: Yale University Press, 2008).

Boyle, J., *Shamans, Software and Spleens: Law and the Construction of the Information Society* (Cambridge and London: Harvard University Press, 1996).

Braithwaite, J. and P. Drahos, *Global Business Regulation* (Cambridge: Cambridge University Press, 2000).

Braybrooke, D., *Studies in Moral Philosophy* (Oxford : Blackwell, 1968).

British Medical Association, *The Medical Profession and Human Rights: Handbook for a Changing Agenda* (London: Zed Books, 2001).

Brown, C., *Sovereignty, Rights and Justice, International Political Theory Today* (Cambridge: Polity Press, 2002).

Brown, W., *States of Injury: Power and Freedom in Late Modernity* (Princeton: Princeton University Press, 1995).

Brownlie, I., *Basic Documents on Human Rights,* 3rd edn (Oxford: Clarendon Press, 1992).

Buchanan, A., *Justice, Legitimacy and Self-Determination: Moral Foundations for International Law* (Oxford: Oxford University Press, 2004).

Buchanan, A. et al., *From Chance to Choice: Genetics and Justice* (Cambridge and New York: Cambridge University Press, 2000).

Chisum, D.S. et al., *Principles of Patent Law: Cases and Materials* (New York: Foundation Press, 2004).

Cho, M.K., *Ethical and Legal Issues of the 21st Century in Preparing for the Millennium: Laboratory Medicine in the 21st Century,* 2nd edn. (Washington, DC: AACC, 1998).

Chossudovsky, M., *The Globalization of Poverty: Impacts of IMF and World Bank Reforms* (London: Zed Books, 1997).

Chossudovsky, M., *The Globalization of Poverty and the New World Order,* 2nd edn. (Shanty Bay: Global Outlook, 2003).

Clapham, A. *Human Rights of Non-State Actors* (New York: Oxford University Press, 2006).

Cohen, W.M. and A. Merrill (eds), *Patents in the Knowledge-Based Economy* (Washington, DC: The National Academies Press, 2003).

Correa, C., *Intellectual Property Rights, the WTO and Developing Countries* (London: Zed Books, 2000).

Cox, R., *Approaches to World Order* (Cambridge: Cambridge University Press, 1996).

Craven, M., *The International Covenant on Economic, Social and Cultural Rights: A Perspective on its Development* (Oxford: Clarendon Press, 1995).

Dahlgren, G. and M. Whitehead, *Policies and Strategies to Promote Social Equality in Health,* (Stockholm: Institute of Future Studies, 1991).

Daniels, N., *Just Health Care* (New York: Cambridge University Press, 1985).

Daniels, N., *Justice and Justification: Reflective Equilibrium in Theory and Practice* (Cambridge: Cambridge University Press, 1996).

David, D.S., *Genetic Dilemmas* (New York: Routledge, 2001).

Davies, G., *Copyright and the Public Interest, Studies in Industrial Property and Copyright Law* (New York: John Wiley & Sons, 1994).

Deepak, N. and J. Court, *Governing Globalization: Issues and Institutions* (Helsinki: United Nations University, 2002).

Descartes, R., *Discourse on Method*, trans. L. Lafleur (New York: Macmillan, 1960).

Dine, J., *The Governance of Corporate Groups* (Cambridge: Cambridge University Press, 2000).

Donnelly, J., *Universal Human Rights in Theory and Practice* (London: Cornell University Press, 1989).

Doyal, L. and I. Gough, *A Theory of Human Need* (New York: The Guildford Press, 1991).

Drahos, P., *A Philosophy of Intellectual Property* (Aldershot: Dartmouth Publishing Company, 1996).

Drahos, P. and R. Mayne (eds), *Global Intellectual Property Rights, Knowledge, Access and Development* (New York: Palgrave Macmillan, 2002).

Dutfield, G., *Intellectual Property Rights and the Life Science Industries: A Twentieth Century History* (Aldershot: Ashgate, 2003).

Engberg-Pedersen, P. et al., *Limits of Adjustment in Africa: The Effects of Economic Liberalization, 1986–94* (Copenhagen: Centre for Development Research, 1996).

Engelhardt Jr, T.H., *The Foundations of Bioethics* (New York: Oxford University Press, 1986).

Evans, T., *US Hegemony and the Project of Universal Human Rights* (London: Macmillan, 1996).

Evans, T. et al., *Challenging Inequities in Health: From Ethics to Action* (New York: Oxford University Press, 2000).

Falk, R., *On Humane Governance: Toward a New Global Politics* (University Park, PA: Pennsylvania State University Press, 1995).

Fletcher, G., *Loyalty: An Essay on the Morality of Relationships* (Oxford: Oxford University Press, 1993).

Feinberg, J., *Social Philosophy* (New Jersey: Prentice Hall, 1973).

Finnis, J., *Natural Law and Natural Rights* (Oxford: Clarendon Press, 1980).

Freemen, M., *Rights* (Minneapolis: University of Minnesota Press, 1991).

French, A.P., *Einstein: a Centenary Volume* (Cambridge: Harvard University Press, 1979).

Fried, C., *Rights and Wrongs* (Cambridge: Harvard University Press, 1978).

Friedman, M., *Capitalism and Freedom* (Chicago: University of Chicago Press, 1962).

Friedman, T., *The Lexus and the Olive Tree* (New York: Farrar Straus Giroux, 1999).

Frost, M., *Ethics in International Relations: A Constitutive Theory* (Cambridge: Cambridge University Press, 1996).

Frost, M., *Towards a Normative Theory of International Relations* (Cambridge: Cambridge University Press, 1986).

Fukuyama, F., *End of History and the Last Man* (New York: Free Press, 1992).

Furet, F., *Marx and the French Revolution* (London: University of Chicago Press, 1995).

Gold, E.R., *Body Parts: From Property Rights to Human Biological Materials* (Washington, DC: Georgetown University Press, 1996).

Goldhaber, M., *Reinventing Technology* (New York: Routledge and Kegan Paul, 1986).

Griffiths, A.J.F. et al., *An Introduction to Genetic Analysis* (New York: W H Freeman & Co, 1999).

Haker, H., R. Hearn and K. Steigleder, *Ethics of the Human Genome Analysis* (Germany: Attempto Verlag Tubingen, 1993).

Hampshire, S., *Innocence and Experience* (Cambridge: Harvard University Press, 1989).

Harris, D., *Justifying State Welfare: the New Right versus the Old Left* (Oxford: B. Blackwell, 1987).

Hegel, G.W., *Hegel's Philosophy of Right,* trans. T.M. Knox (Oxford: Clarendon Press, 1964).
Heller, A., *The Theory of Human Need in Marx* (London: Allison & Busby, 1976).
Herbert, G.B., *A Philosophical History of Rights* (London: Transaction Publishers, 2002).
Hestermeyer, H. *Human Rights and the WTO: The Case of Patents and Access to Medicines* (New York: Oxford University Press, 2007).
Hohfeld, W., *Fundamental Legal Conceptions* (New Haven: Yale University Press, 1923 and Aldershot: Ashgate, 2001).
Hoogvelt, A., *Globalization and the Postcolonial World: The New Political Economy of Development* (Basingstoke: Palgrave, 2001).
Hymer, S., *The Multinational Corporation: a Radical Approach* (Cambrige: Cambridge University Press, 1979).
Jawara, F. and A. Kwa, *Behind the Scenes at The WTO: The Real World of International Trade Negotiations* (London and New York: Zed Books, 2003).
Jewkes, J., D. Sawers and R.Stillerman, *The Sources of Invention*, 2nd edn. (New York: W.W. Norton, 1969).
Jones, C., *Global Justice, Defending Cosmopolitanism* (Oxford: Oxford University Press, 1999).
Jones, P., *Rights* (London: Macmillan, 1994).
Jones, S., *Genetics in Medicine: Real Promises, Unreal Expectations. One Scientist's Advice to Policymakers in the United Kingdom and the United States* (London: Milbank Memorial Fund, 2000).
Kameri-Mbote, P., *Property Rights and Biodiversity Management in Kenya* (Act Press: Kenya, 2002).
Kant, I., *Foundations of the Metaphysics of Morals,* 2nd edn., trans. L. White Beck (London: Collier Macmillan, 1990).
Kant, I., *The Metaphysical Elements of Justice (*1797), trans. J. Ladd, 2nd edn. (Indianapolis: Hackett Publishing Co., 1999).
Kaul, I., *Providing Global Public Goods: Managing Globalization* (New York: Oxford University Press, 2003).
Keck, M.E. and K. Sikkink, *Activists Beyond Borders: Advocacy Networks in International Politics* (Ithaca: Cornell University Press, 1998).
Kymlicka, W., *Contemporary Political Philosophy: An Introduction*, 2nd edn. (Oxford: Oxford University Press, 2002).
Kymlicka, W., *Liberalism, Community and Culture* (Oxford: Clarendon Press, 1989).
Ladas, S., *Patents, Trademarks and Related Rights – National and International Protection* (Cambridge: Harvard University Press, 1975).
Lederer, K., *Human Needs* (Cambridge: Oelgeschlager, Gunn & Hain, 1980).

Lerner, M.J., *The Belief in a Just World: A Fundamental Delusion* (New York: Plenum Press, 1980).

Linklater, A., *Transformation of Political Community* (London: Polity Press, 1998).

Locke, J., *Second Treatise of Government* in P. Laslett (ed) (Cambridge: Cambridge University Press, 1988).

Locke, J., *Two Treatises of Government* in P. Laslett (ed) (Cambridge: Cambridge University Press, 1960) (originally published in 1689).

Loewenson, R. and A. Whiteside, *United Nations Development Programme, HIV/AIDS: Implications for Poverty Reduction* (New York: United Nations Development Programme, 2001).

Machlup, F., *Production and Distribution of Knowledge in the United States* (Princeton: Princeton University Press, 1962).

Macpherson, C.B., *Democratic Theory: Essays in Retrieval* (Oxford: Clarendon Press, 1973).

Mandel, M., *The Charter of Rights & the Legalization of Politics in Canada* (Toronto: Thompson Educational Publishing, 1994).

Mandelbaum, M., *The Ideas that Conquered the World* (New York: Public Affairs, 2002).

Mann, J. et al., *Health and Human Rights* (New York: Routledge, 1999).

Martin, J., *Globalization and Development: A Latin American and Caribbean Perspective* (Palo Alto: Stanford University Press, 2003).

Martin, R., *A System of Rights* (Oxford: Clarendon, 1993).

Marx, K. and F. Engels, *Selected Works, Vol. Three, 1875–1895* (Moscow: Foreign Languages Publishing House, 1970).

Maskus, K., *Intellectual Property Rights in the Global Economy* (Washington, DC: Institute for International Economics, 2000).

May, C., *A Global Political Economy of Intellectual Property Rights: The New Enclosures?* (London: Routledge, 2000).

Mehlman, M.J. and J.R. Botkin, *Access to the Genome: The Challenge to Equality* (Washington, DC: Georgetown University Press, 1998).

Melucci, A., *Nomads of the Present: Social Movements and Individual Needs in Contemporary Society* (Philadelphia: Temple University Press, 1989).

Morgenthau, H.J., *Politics Among Nations: The Struggle for Power and Peace,* 6th edn. (New York: Alfred A. Knopf, 1985).

Mower, G.A., *International Cooperation for Social Justice: Global and Regional Protection of Economic/Social Rights* (Westport: Greenwood Press, 1985).

Muchlinski, P.T., *Multinational Enterprises and the Law* (Oxford: Blackwell Publishers, 1995).

Mutua, M., *Human Rights: A Political and Cultural Critique* (Philadelphia: University of Pennsylvania Press, 2002).

Myrdal, G., *Beyond the Welfare State* (New Haven: Yale University Press, 1960).

Nagel, T., *Mortal Questions* (Cambridge: Cambridge University Press, 1979).

Narveson, J., *The Libertarian Idea* (Philadelphia: Temple University Press, 1988).

Noble, D., *America by Design* (New York: Knopf, 1982).

Nozick, R., *Anarchy, State and Utopia* (Oxford: Blackwell, 1974).

O'Neill, O., *Bounds of Justice* (Cambridge: Cambridge University Press, 2000).

O'Neill, O., *Face of Hunger: An Essay in Poverty, Development and Justice* (London: Allen and Unwin, 1986).

O'Neill, O., *Towards Justice and Virtue* (Cambridge: Cambridge University Press, 1996).

Oslon, M., *The Logic of Collective Action* (Cambridge: Cambridge University Press, 1965).

Parkinson, J.E., *Corporate Power and Responsibility* (Oxford: Clarendon Press, 1993).

Payne, R.A. and N.H. Samhat, *Democratizing Global Politics, Discourse Norms, International Regimes and Political Community* (New York: State University of New York Press, 2004).

Penrose, E., *The Economics of the International Patent System* (Baltimore, MD: Johns Hopkins Press, 1951).

Peters, T., *Playing God: Genetic Determinism and Human Freedom* (New York: Routledge, 1997).

Plant, R. and N. Barry, *Citizenship and Rights in Thatcher's Britain: Two Views* (London: IEA Health & Welfare Unit, 1992).

Pogge, T.W., *Realizing Rawls* (Ithaca: Cornell University Press, 1989).

Pogge, T.W., *Recognized and Violated by International Law: The Human Rights of the Global Poor* (forthcoming).

Pogge, T.W., *World Poverty and Human Rights: Cosmopolitan Responsibilities and Reforms* (Cambridge: Polity Press, 2002).

Rawls, J., *A Theory of Justice* (Cambridge: Harvard University Press, 1971).

Rawls, J., *A Theory of Justice*, 2nd edn. (Oxford: Oxford University Press, 1999).

Rawls, J., *The Law of Peoples* (Cambridge: Harvard University Press, 1999).

Raz, J., *The Morality of Freedom* (Oxford: Clarendon Press, 1986).

Resnik, D.B., *Owning the Genome: a Moral analysis of DNA Patenting* (Albany: State University of New York Press, 2003).

Richards, D.G., *Intellectual Property Rights and Global Capitalism, The Political Economy of the TRIPS Agreement* (London: M.E. Sharpe, 2004).

Rittich, K., *Recharacterizing Restructuring: Law, Distribution and Gender in Market Reform* (The Hague: Kluwer Law International, 2002).

Robinson, J., *Prescription Games: Money, Ego, and Power Inside the Global Pharmaceutical Industry* (London: Simon & Schuster, 2001).

Roemer, J., *Equality of Opportunity (*Cambridge: Cambridge University Press, 1998).

Rolston, H., *Conserving Natural Value* (New York: Columbia University Press, 1994).

Rose, H., *The Commodification of Bioinformation: The Icelandic Health Sector Database* (London: The Wellcome Trust, 2001).

Ryan, M., *Knowledge Diplomacy: Global Competition and the Politics of Intellectual Property* (Washington DC, Brookings Institution, 1998).

Sach, J., *The End of Poverty: Economic Possibilities for Our Time* (New York: Penguin Press, 2005).

Sherman, B. and L. Bently, *The Making of Modern Intellectual Property Law* (Cambridge: Cambridge University Press, 1999).

Shulman, S., *Owning the Future* (Boston: Houghton Mifflin, 1999).

Sell, S.K., *Private Power, Public Law: The Globalization of Intellectual Property Rights*, (Cambridge: Cambridge University Press, 2003).

Sen, A., *Commodities and Capabilities* (New York: Elvier Science Publishing, 1985).

Sen, A., *Development as Freedom* (New York: Knopf, 1999).

Sen, A., *The Standard of Living, The Tanner Lectures* (Cambridge: Cambridge University Press, 1987).

Shue, H., *Basic Rights: Subsistence, Affluence and US Foreign Policy,* 2nd edn. (Princeton: Princeton University Press, 1996).

Shute, S. and S. Hurley (eds), On *Human Rights: The Oxford Amnesty Lecture* (New York: Basic Books, 1993).

Singer, J.W., *Entitlement: The Paradoxes of Property* (New Haven: Yale University Press, 2000).

Smith, A., *Adam Smith's Wealth of Nations: a New and Condensed Edition* (New York: T.Y. Crowell, 1904).

Stiglitz, J.E., *Globalization and its Discontents* (New York: WW Norton & Co. inc., 2002).

Stiglitz, J.E. and A. Charlton, *Fair Trade for All: How Trade can Promote Development* (New York: Oxford University Press, 2005).

Stine, G., *AIDS Update 2000* (Upper Saddle River, NJ: Prentice Hall, 2000).

St-Onge, J.-C., *L'envers de la Pilule* (Montréal: Ecosociété, 2004).

Taylor, C., *Philosophy and the Human Sciences: Philosophical Papers 2* (Cambridge: Cambridge University Press, 1985).

Taylor, C., *Reconciling the Solitudes* (Kingston and Montreal: McGill-Queens, 1993).

Terence, B. and R. Dagger, *Political Ideologies and the Democratic Ideal*, 5th edn. (New York: Addison Wesley, 2004).

Thomas, C.G., *Vulnerable Subjects: Ethics and Life Writing* (Ithaca: Cornell University Press, 2004).

Trebilcock, M. and R. Howse, *The Regulation of International Trade* (New York: Routledge, 1995).

Vincent, R.J., *Human Rights and International Relations* (Cambridge: Cambridge University Press, 1998).

Waldron, J., *Liberal Rights: Collected Papers* (Cambridge: Cambridge University Press, 1993).

Waldron, J., *Nonsense upon Stilts: Bentham, Burke and Marx on the Rights of Man* (London: Methuen, 1987).

Waldron, J., *The Right to Private Property* (Oxford: Clarendon Press, 1988).

Walzer, M., *Spheres of Justice* (Oxford: Blackwell, 1983).

Wellman, C., *An Approach to Rights: Studies in the Philosophy of Law and Morals* (Boston: Kluwer Publishers, 1997).

Wellman, C., *The Proliferation of Rights: Moral Progress or Empty Rhetoric* (Boulder: Westview Press, 1999).

Wilson, R.A., *Human Rights, Culture and Context* (London: Pluto Press, 1997).

Woods, A., *Hegel's Ethical Thought* (New York: Cambridge University Press, 1990).

Zakaria, F., *The Future of Freedom* (New York: Norton, 2003).

SECONDARY MATERIAL: COLLECTIONS OF ESSAYS

Akermark, S.S. 'International Development Finance Institutions: the World Bank and the International Monetary Fund Agencies' in A. Eide et al. (eds), *Economic, Social and Cultural Rights: A Textbook*, 2nd edn (London: Maartinus Nijhoff, 2001).

Alper, J., 'Genetic Complexity in Human Disease and Behavior' in J. Alper et al. (eds), *The Double-Edged Helix: Social Implications of Genetics in a Diverse World* (Baltimore: Johns Hopkins University Press, 2002).

Annas, G.J., 'The Impact of Health Policies on Human Rights: AIDS and TB Control' in J.M. Mann et al. (eds), *Health and Human Rights: A Reader* (New York: Routledge, 1999).

Aoki, K. 'The Stakes of Intellectual Property Law' in D. Kairys (ed)., *The Politics of Law: A Progressive Critique* (New York: Basic Books, 1998).

Arthur, J., 'Rights and the Duty to Bring Aid' in W. Aiken and H. LaFolette (eds), *World Hunger and Morality,* 2nd edn. (New York: Prentice-Hall, 1996).

Balasubramaniam, K., 'Access to Medicine: Patents, Price and Public Policy – Consumer Perspective' in P. Drahos and R. Mayne (eds), *Global*

Intellectual Property Rights, Knowledge, Access and Development (New York: Palgrave Macmillan, 2002).

Bard-Anders, A. et al., 'Assessing Human Rights Performance in Developing Countries: The Case for a Minimal Threshold Approach to Economic and Social Rights' in A. Bard-Anders and E. Asbjorn (eds), *Human Rights in Developing Countries* (Copenhagen: Akademisk Forlag. 1988).

Barton, J.H., 'Intellectual Property Rights and Innovation' in N. Imparato (ed), *Capital for Our Time: the Economic, Legal, and Management Challenges of Intellectual Capital* (Stanford: Hoover Institution Press, 1999).

Beitz, C.R., 'Cosmopolitan Liberalism and the States System' in C. Brown (ed), *Political Restructuring in Europe, Ethical Perspectives* (NewYork: Routledge, 1994).

Beitz, C.R., 'International Justice: Conflict' in L.C. Becker and C. Becker (eds), *Encyclopedia of Ethics* (London: Garland, 1992).

Benatar, S.R., 'A Perspective from Africa on Human Rights and Genetic Engineering' in R. Dawkins (ed), *The Genetic Revolution and Human Rights* (Oxford: Oxford University Press, 1999).

Bentham, J., 'The Theory of Legislation' in C.B. Macpherson (ed), *Property, Mainstream and Critical Positions* (Toronto: University of Toronto Press, 1978).

Booth, K., 'Three Tyrannies' in T. Dunne and N. Wheeler (eds), *Human Rights in Global Politics* (New York: Cambridge University Press, 1999).

Brown, C., 'Universal Human Rights: A Critique' in T. Dunne and N.J. Wheeler (eds), *Human Rights in Global Politics* (Cambridge: Cambridge University Press, 1999).

Bubela, T.M. and T. Caulfield, 'Media Representations of Genetic Research' in E.F. Einsiedel and F. Timmermans (eds), *Crossing Over. Genomics in the Public Arena* (Calgary: University of Calgary Press, 2005).

Carens, J., 'Migration and Mortality: A Liberal Egalitarian Perspective' in B. Barry and R.E. Goodin (eds), *Free Movement* (Hemel Hempstead: Harvester Wheatsheaf, 1992).

Chadwick, A.F., 'Transnational Social Movements, World Politics and Global Governance' in J. Smith, C. Chatfield and R. Pugnucco, *Transnational Social Movements and Global Politics: Solidarity Beyond the State* (Syracuse, NY: Syracuse University Press, 1997).

Chan, J., 'A Confucian Perspective on Human Rights for Contemporary China' in J.R. Bauer and D.A. Bell (eds), *The East Asian Challenge for Human Rights* (Cambridge: Cambridge University Press, 1999).

Chapman, A., 'Reintegrating Right and Responsibilities' in K.W. Hunter and T.C. Mack (eds), *International Rights and Responsibilities for the Future* (Westport, CT: Praeger, 1996).

Child, J.W., 'The Moral Foundations of Intangible Property' in A.D. Moore (ed), *Intellectual Property: Moral, Legal, and International Dilemmas* (Maryland: Rowman & Littlefield, Lanham, 1997).

Coomaraswamy, R., 'Reinventing International Law: Women's Rights as Human Rights in the International Community' in P. Van Ness (ed), *Debating Human Rights: Critical Essays from the United States and Asia* (London: Routledge, 1999).

Daniels, N., 'The Genome Project: Individual Differences and Just Health Care' in: T.F. Murphy and Marc A. Lappé (eds), *Justice and the Human Genome Project* (California: University of California Press, 1994).

David, P., 'Intellectual Property Institutions and the Panda's Thumb: Patents, Copyrights, and Trade Secrets in Economic Theory and History' in M.B. Wallerstein, M.E. Mogee and R.A. Schoen (eds), *National Research Council, Global Dimensions of Intellectual Property Rights In Science and Technology* (New York: National Academy Press, 1993).

Donnelly, J., 'Human Rights and Asian Values: A Defense of "Western Universalism" ' in J.R. Bauer and D.A. Bell (eds), *The East Asian Challenge for Human Rights* (Cambridge: Cambridge University Press, 1999).

Donnelly, J., 'The Social Construction of International Human Rights' in T. Dunne and N.J. Wheeler (eds), *Human Rights in Global Politics* (Cambridge: Cambridge University Press, 1999).

Drahos, P., 'The Rights to Food and Health and Intellectual Property in the Era of "Biogopolies"' in S. Bottomley and D. Kinley (eds), *Commercial Law and Human Rights* (Aldershot: Ashgate Dartmouth, 2002).

Eide, A., 'Economic, Social and Cultural Rights as Human Rights' in A. Eide et al. (eds), *Economic, Social and Cultural Rights: A Texbook*, 2nd edn. (London: Maartinus Nijhoff, 2001).

Eide, A., 'Future Protection of Economic and Social Rights in Europe' in A. Bloed et al. (eds), *Monitoring Human Rights in Europe: Comparing International Procedure and Mechanisms* (Boston: Maartinus Nijhoff, 1993).

Farmer, P., 'On Suffering and Social Violence: A View from Below' in A. Kleinman, V. Das and M. Lock (eds), *Social Suffering* (Berkeley: University of California Press, 1997).

Gerwirth, A., 'Rights' in L.C. Becker and C. Becker (eds), *Encyclopedia of Ethics* (London: Garland, 1992).

Gutmann, A., 'For and Against Equal Access to Health Care' in President's Commission for the Study of Ethical Problems in Medicine and Biomedical and Behavioral Research (ed) *Securing Access to Health Care* (Washington, DC: US Government Printing Office, 1983).

Harrison, S. and M. Qose, 'Health Legislation' in A. Ntuli (ed), *South African Health Review 1998* (Durban: Health Systems Trust, 1999).

Hettinger, E.C., 'Justifying Intellectual Property' in A.D. Moore (ed), *Intellec-*

tual Property: Moral, Legal, and International Dilemmas (Maryland: Rowman & Littlefield, Lanham, 1997).

Hettinger, E.C., 'Justifying Intellectual Property' in P. Drahos (ed), *Intellectual Property* (Aldershot: Dartmouth Publishing, 1999).

Hossain, K., 'Globalisation and Human Rights: Clash of Universal Aspirations and Special Interests' in E. Burns, H. Weston and S.P Marks, *The Future of International Human Rights*, (Transnational Publishers, New York, 1999).

Kardos, G., 'Right to Peace, Right to Development, Right to a Healthy Environment: Part of the Solution or Part of the Problem' in A. Rosas and J. Helgesen (eds), *Human Rights in a Changing East-West Perspective* (New York: Pinter Publishers, 1990).

Kieff, S.F., 'Perusing Property Rights in DNA' in S.F. Kieff (ed), *Perspectives on Properties of the Human Genome Project* (California: Elsevier, 2003).

Krause, C., 'The Right to Property' in A. Eide et al. (eds), *Economic, Social and Cultural Rights: A Texbook* 2nd edn. (London: Maartinus Nijhoff, 2001).

Laertius, D., 'Diogenes' in *Lives of Eminent Philosophers*, trans. R.D. Hicks, Loeb Classical Library (London: William Heinemann, 1925).

Leckie, S., 'Violations of Economic, Social and Cultural Rights' in T.C. van Boven, C. Flinterman and I. Westendorp (eds), *The Maastricht Guidelines on Violations of Economic, Social and Cultural Rights – SIM Special No. 20* (Utrecht: 1998).

Leonard, M., 'Just Genetics: A Problem Agenda' in T.F. Murphy and M.A. Lappé (eds), *Justice and the Human Genome Project* (California: University of California Press, 1994).

Leonard, M., 'Violations of Socio-Economic Rights: The Role of the South African Human Rights Commission' in P. Andrews and S. Ellmann (eds), *The Post-Apartheid Constitutions: Perspectives on South Africa's Basic Law* (Johannesburg: Witwatersrand University Press, 2001).

Love, J., 'Access to Medicine and Compliance with WTO/TRIPS Accord' in P. Drahos and R. Mayne (eds), *Global Intellectual Property Rights, Knowledge, Access and Development* (New York: Palgrave Macmillan, 2002).

MacDonald, M., 'Natural rights' in J. Waldron (ed), *Theories of Rights* (Oxford: Oxford University Press, 1984).

Macdonald, S., 'Exploring the hidden costs of patents' in P. Drahos and R. Mayne (eds), *Global Intellectual Property Rights, Knowledge, Access and Development* (New York: Palgrave Macmillan, 2002).

Mann, J., 'Human Rights and AIDS: The Future of the Pandemic' in J.M. Mann et al. (eds), *Health and Human Rights* (New York: Routledge, 1999).

Martin, B., 'Against Intellectual Property' in P. Drahos (ed), *Intellectual Property* (Aldershot: Dartmouth Publishing, 1999).

Marx, K., 'On the Jewish Question' in J. O'Walley (ed), *Early Political Writing* (Cambridge: Cambridge University Press, 1994).

Marx, K., 'Capital' in F. Engels (ed), *Manisfesto of the Communist Party,* Translated from the 3rd German edn. by S. Moore and E. Aveling. Rev. (Chicago: Encyclopædia Britannica, 1955).

McEvoy, A. 'Market and Ethics in United States Property Law' in H.M. Jacobs (ed), *Who Owns America? Social Conflict Over Property Rights* (Madison: University of Wisconsin Press, 1998).

Meintjes, G., 'An International Human Rights Perspective on Corporate Codes' in O.F. Williams (ed), *Global Codes of Conduct: An Idea Whose Time has Come* (Indiana: University of Notre Dame Press, 2000).

Miller, D., 'The Nation-State: a Modest Defence' in C. Brown (ed), *Political Restructuring in Europe, Ethical Perspectives* (NewYork: Routledge, 1994).

Moore, A.D., 'Toward a Lockean Theory of Intellectual Property' in A.D. Moore, *Intellectual Property: Moral, Legal, and International Dilemmas* (Maryland: Rowman & Littlefield, 1997).

Nagel, T., 'Poverty and Food: Why Charity Is Not Enough' in P.G. Brown and H. Shue (eds), *Food Policy* (New York: The Free Press, 1977).

Nielsen, K., 'Global Justice, Capitalism, and the Third World' in John Arthur and William H. Shaw (ed), *Justice and Economic Distribution* (Englewood Cliffs: Prentice Hall, 2001).

O'Neill, O., 'Hunger, Needs and Rights' in S. Luper-Foy (ed), *Problems of International Justice* (London: Westview, 1988).

O'Neill, O., 'Justice and Boundaries' in C. Brown (ed), *Political Restructuring in Europe: Ethical Perspectives* (London: Routledge, 1994).

O'Neill, O., 'Justice Gender and International Boundaries' in M.C. Nussbaum and Amartya Sen (eds), *Quality of Life* (Oxford: Clarendon, 1993).

Oyugi, W.O., 'Service Provision in Rural Kenya: Who Benefits?' in J. Semboja and O. Therkeldsen (eds), *Service Provision under Stress in East Africa* (Copenhagen: Centre for Development Research, 1995).

Parekh, B., 'Non-Ethnocentric Universalism' in T. Dunne and N.J. Wheeler (eds), *Human Rights in Global Politics* (Cambridge: Cambridge University Press, 1999).

Peters, F. and T. Evans, 'Ethical Dimensions of Health Equity' in T. Evans et al. (eds), *Challenging Inequities in Health: From Ethics to Action* (New York: Oxford University Press, 2001).

Petersmann, E.-U., 'National Constitutions and International Economic Law' in M. Hilf and E.-U. Petersmann (eds), *National Constitutions and International Economic Law* (Deventer: Kluwer, 1993).

Petersmann, E.-U., 'Democratizing Globalism' in D. Drache (ed), *The Market or the Public Domain, Global Governance and the Asymmetry of Power* (New York: Routledge, 2001).

Pogge, T.W., 'A Global Resources Dividend' in D. Crocker and T. Linden (eds), *Ethics of Consumption* (Totowa, NJ: Roman and Littlefield, 1999).

Pretorius, W., 'TRIPS and Developing Countries: How Level is the Playing Field?' in P. Drahos and R. Mayne (eds), *Global Intellectual Property Rights, Knowledge, Access and Development* (New York: Palgrave Macmillan, 2002).

Rakowski, E., 'Cosmopolitanism and Sovereignty' in C. Brown (ed), *Political Restructuring in Europe, Ethical Perspectives* (NewYork: Routledge, 1994).

Rorty, R., 'Human Rights, Rationality, and Sentimentality' in S. Shute and S. Hurley (eds), *On Human Rights: the Oxford Amnesty Lectures 1993* (New York: Basic Books, 1993).

Rosas, A., 'The Right to Development' in A. Eide et al. (eds), *Economic, Social and Cultural Rights: A Texbook*, 2nd edn. (London: Maartinus Nijhoff, 2001).

Scheffler, S. 'Liberalism, Nationalism, and Egalitarianism' in R. McKim and J. McMahan (eds), *The Morality of Nationalism* (New York: Oxford University Press, 1997).

Schrecker, T., 'Benefit-Sharing in the New Genomic Marketplace: Expanding the Ethical Frame of Reference' in B.M. Knoppers (ed), *Populations and Genetics: Legal and Socio-Ethical Perspectives* (Leiden: Martinus Nijhoff, 2003).

Spector, H.M, 'An outline of a Theory Justifying Intellectual and Industrial Property Rights' in P. Drahos (ed), *Intellectual Property* (Aldershot: Dartmouth Publishing, 1999).

Sulston, J., 'IP and the Human Genome' in P. Drahos and R. Mayne (eds), *Global Intellectual Property Rights, Knowledge, Access and Development* (New York: Palgrave Macmillan, 2002).

Thomas, C.G., 'International Financial Institutions and Social and Economic Human Rights: An Exploration' in T. Evans (ed), *Human Rights Fifty Years On: A Reappraisal* (Manchester: Manchester University Press, 1998).

Thompson, J., 'Community, Identity and World Citizenship' in D. Archibugi, D. Held and M. Köhler (eds), *Re-imagining Political Community: Studies in Cosmopolitan Democracy* (Stanford, CA: Stanford University Press, 1998).

Toebes, B., 'The Right to Health' in A. Eide et al. (eds), *Economic, Social and Cultural Rights: A Textbook*, 2nd edn. (London: Maartinus Nijhoff, 2001).

van Rensburg, D. et al., 'Primary Health Care Facilities Survey' in A. Ntuli et al. (eds), *South African Health Review 2000* (Durban: Health Systems Trust, 2001).

Vaver, D., 'Intellectual Property Today: Of Myths and Paradoxes' in Drahos, P. (ed), *Intellectual Property* (Aldershot: Dartmouth Publishing, 1999).

Walsh, J.P., A. Arora and W.M. Cohen, 'Effects of Research Tool Patents and Licensing on Biomedical Innovation' in W.M. Cohen and A. Merrill (eds),

Patents in the Knowledge-Based Economy (Washington, DC: The National Academies Press, 2003) 285, online on the NAP website: http://books.nap.edu/books/0309086361/html/297.html#pagetop (accessed 26 May 2006).

Williams, B., 'A Critique of Utilitarianism' in J.J.C. Smart (ed), *Utilitarianism: For and Against* (Cambridge: Cambridge University Press, 1973).

SECONDARY MATERIAL: ARTICLES

Abott, A., 'DNA Study Deepens Rift Over Iceland's Genetic Heritage' (2003) 421: 6924 *Nature* 678.

Abbott, F., 'The Enduring Enigma of TRIPs' (1998) 1 *Journal of International Economic Law* 506.

Acharya, A.K., 'Toward Establishing a Universal Basic Health Norm' (2004) 18:3 *Ethics and International Affairs* 65.

Adewole, T.A. et al., 'Application of Polymerase Chain Reaction to the Prenatal Diagnosis of Sickle Cell Anaemia in Nigeria' (July–Sep 1999) 18:3 *West Afr J. Med.* 160.

Adler, R., 'Genome Research: Fulfilling the Public's Expectations for Knowledge and Commercialization' (1992) 257 *Science* 908.

Agbacka, S.A., 'Reclaiming Humanity: Economic, Social and Cultural Rights as the Cornerstone of African Human Rights' (Annual 2002) 5 *Yale Human Rights and Development Law Journal* 177.

Akinyanju, O.O. et al., 'Initiation of Prenatal Diagnosis of Sickle-cell Disorders' (1999) 19 *Prenat. Diagn.* 299.

Albert, B. and A. Klug, 'The Human Genome Itself Must be Freely Available to all Humankind' (2000) 404 *Nature* 325.

Allhoff, F., 'Germ-Line Genetic Enhancement and Rawlsian Primary Goods' (May 2008) 18:1 *Journal of Evolution and Technology* 10.

Alston, P., 'The Myopia of the Handmaidens: International Lawyers and Globalization' (1997) 8 *Eur. J. Int'l L.* 435.

Alston, P., 'Out of the Abyss: The Challenge Confronting the New UN Committee on Economic, Social and Cultural Rights' (1987) 9 *Human Rights Quarterly* 332.

Alston, P., 'A Third Generation of Solidarity Rights: Progressive Development or Obfuscation of International HR law' (1982) 29:3 *Netherlands International Law Review* 307.

Altmuller, J. et al., 'Genomewide Scans of Complex Human Diseases: True Linkage is Hard to Find' (2001) 69 *American Journal of Human Genetics* 936.

Alwan, A. and B. Modell, 'Opinion: Recommendations for Introducing Genetics Services in Developing Countries' (2003) 4:1 *Nat Rev Genet.* 61.

Amason, E., 'Genetic Heterogeneity of Icelanders' (2003) 67:1 *Ann Hum Genet.* 5.

Arnason, V., 'Coding and Consent: Moral Challenges of the Database Project in Iceland' (2004) 18:1 *Bioethics* 27.

American Medical Association, Council on Ethical and Judicial Affairs, 'Ethical Issues Related to Prenatal Genetic Testing' (1994) 3 *Archives of Family Medicine* 633.

American Society of Human Genetics, 'Professional Disclosure of Familial Genetic Information' (1998) 62:2 *American Journal of Human Genetics* 474.

Anderson, E., 'What is the Point of Equality?' (1999) 109:2 *Ethics* 287.

Andrews, L.B., 'Past as Prologue: Sobering Thoughts on Genetic Enthusiasm' (1997) 27 *Seton Hall L. Rev.* 893.

Angell, M., 'The Pharmaceutical Industry: To Whom is it Accountable?' (2000) 342:25 *New Eng. J. Med.* 1902.

Anghie, A., 'Time Present and Time Past: Globalization, International Financial Institutions, and the Third World' (2000) 32 *N.Y.U. J. Int'l L. & Pol.* 243.

Anand, S., 'The Concern for Equity in Health' (2002) 12:1 Harvard Center for Population and Development Studies Working Paper Series.

Anaya, S.J. and R.A. Williams, Jr., 'The Protection of Indigenous Peoples' Rights over Lands and Natural Resources Under the Inter-American Human Rights System' (2001) 14 *Harvard Human Rights Journal* 33.

Andrews, L.B., 'The Gene Patent Dilemma: Balancing Commercial Incentives with Health Needs' (2002) 2 *Hous. J. Health & Pol'y* 65.

Annas, G.J., 'The Right to Health and the Nevirapine Case in South Africa' (2003) 348:8 *N. Engl. J. Med.* 750.

Aoki, K., '(Intellectual) Property and Sovereignty: Notes Toward a Cultural Geography of Authorship' (May 1996) 48:5 *Stanford Law Review* 1293.

Aoki, K., 'Neo-colonialism, Anticommons Property and Biopiracy in the (Not-So-Brave) New World Order of International Intellectual Property Protection' (1998) 6 *Ind. J. Global Leg. Stud.* 11.

Arnold, J. and N. Hilton, 'Genome Sequencing: Revelations from a Bread Mould' (2003) 422 *Nature* 821.

Arrow, K., 'The Economics of Information: An Exposition' (1996) 23:2 *Empirica* 125.

Arup, C., 'Competition over Competition Policy for International Trade and Intellectual Property' (1998) 16:3 *Prometheus* 367.

Arup, C., 'How Do Patents and Economic Policies Affect Access to Essential Medicines in Developing Countries?' (2004) 23:3 *Health Affairs* 55.

Arup, C., 'Human Rights and Biomedical Research Funding for The Developing World: Covering State Obligations Under the Right to Health' (1999) 4:1 *Health Hum Rights* 26.

Arup, C. and L. Gillespie-White, 'Do Patents for Antiretroviral Drugs Constrain Access to AIDS Treatment in Africa? (2001) 286:15 *JAMA* 1886.

Austin, W., 'Using the Human Rights Paradigm in Health Ethics: The Problems and the Possibilities' (2001) 8:3 *Nursing Ethics* 183.

Bale, H.E., 'The Conflict Between Parallel Trade and Product Access and Innovation: The Case of Pharmaceuticals' (1998) 1:4 *Journal of International Economic Law* 637.

Baleta, A., 'Africa Soaks up Pressure to Change HIV/AIDS policy' (2002) *Lancet* 360.

Baleta, A., 'Drug Firms Lose Patent Rights Lawsuit Against South Africa's Government' (2001) 357 *Lancet* 1347.

Ball, P., 'Synthetic Biology: Starting from Scratch' (2004) 431 *Nature* 624.

Barry, C. and K. Raworth, 'Access to Medicines and the Rhetoric of Responsibility' (2002) 16:2 *Ethics and International Affairs* 57.

Barton, J.H. and E.J. Emmanuel, 'The Patents-Based Pharmaceutical Development Process: Rationale, Problems and the Potential Reforms' (2005) 294:16 *The Journal of the American Medical Association* 2075.

Bass, N.A., 'Implications of the TRIPS Agreement for Developing Countries: Pharmaceutical Patent Laws in Brazil and South Africa in the 21st Century' (2002) 34:1 *George Washington International Law Review* 191.

Basu, P., 'International Patent Law – Boon or Bane of Biotech?' (2005) 23:1 *Nature Biotechnology* 13.

Becker, L., 'The Labour Theory of Property Acquisition' (1976) *Journal of Philosophy* 653.

Beier, F.-K., 'Exclusive Rights, Statutory Licenses and Compulsory Licenses in Patent and Utility Model Law' (1999) 30:3 *International Review of Industrial Property and Copyright Law* 251.

Beier, F.-K. and J. Straus, 'The Patent System and its Informational Function – Yesterday and Today' (1977) 8 *Int'l Rev. Int. Prop. & C'right L.* 387.

Beitz, C.R., 'Human Rights as a Common Concern' (2001) 95:2 *American Political Science Review* 269.

Beitz, C.R., 'International Liberalism and Distributive Justice: A Survey of Recent Thought' (1999) 51:2 *World Politics* 269.

Beitz, C.R., 'Rawls's Law of Peoples' (July 2000) 110 *Ethics* 669.

Beitz, C.R., 'Social and Cosmopolitanism Liberalism' (1999) 75:3 *International Affairs* 515.

Béland, F., 'The Supreme Court Missed a Good Opportunity' (2005) *Law and Governance*.

Bell, J., 'The Double Helix in Clinical Practice' (2003) 421 *Nature* 414.

Benatar, S.R., A.S. Daar and P. Singer, 'Global Health Ethics: The Rationale for Mutual Caring' (2003) 79:1 *International Affairs* 107.

Bihai, B.E. et al., 'Génomique, Promesses et Réalités' (2000) 16:1 *Médecine/Sciences* 17.

Bilchitz, D., 'Placing Basic Needs at the Centre of Socio-Economic Rights Jurisprudence' (2003) 4:1 *Socio-Economic Rights Project.*

Bilchitz, D., 'South Africa: Right to Health and Access to HIV/AIDS Drug Treatment' (2003) 1:3 *International Journal of Constitutional Law* 524.

Blake, M., 'Distributive Justice, Sate Coercion and Autonomy' (2001) 3 *Philosophy & Public Affairs* 257.

Bloom, B.R. and D.D. Trach, 'Genetics and Developing Countries' (2001) 322:7293 *British Medical Journal* 1006.

Boetzkes, E., 'Genetic Knowledge and Third-Party Interests' (1999) 8:4 *Cambridge Quarterly of Healthcare Ethics* 386.

Bojang, K.A. et al., 'Efficacy of RTS,S/ASO2 Malaria Vaccine Against *Plasmodium Falciparum* Infection in Semi-Immune Adult Men in The Gambia: a Randomised Trial' (2001) 358 *Lancet* 1927.

Boorse, C., 'Health as a Theoretical Concept' (1977) 44 *Philosophy of Science* 542.

Boorse, C., 'On the Distinction between Disease and Illness' (1975) 5:1 *Philosophy & Public Affairs* 49.

Boulier, W., 'Sperm, Spleens and Other Valuables: The Need to Recognize Property Rights in Human Body Parts' (1995) 23 *Hofstra L. Rev* 705.

Bourgain, C. et al., 'Search for Multifactorial Disease Suceptibility Genes in Founder Populations' (2000) 64 *Annals Human Genetics* 255.

Bovenberg, J., 'Whose Tissue is it Anyway?' (2005) 23:8 *Nature Biotechnology* 929.

Boylan, M., 'Medical Pharmaceuticals and Distributive Justice' (2008), 17 *Cambridge Quarterly of Healthcare Ethics* 30.

Boyle, J., 'A Politics of Intellectual Property: Environmentalism for the Net?" (1997) 47:1 *Duke Law Journal* 89.

Boyle, J., 'A Theory of Law and Information: Copyright, Spleens, Blackmail, and Insider Trading' (1992) 80 *California Law Review* 1415.

Braude, P. et al., 'Preimplantation Genetic Diagnosis' (2002) 3:12 *Nature Review Genetics* 941.

Brower, V., 'Canavan Families Slam Scientists over Test Patent Profits' (2000) *Biotechnology Newswatch* 1.

Brower, V., 'Tackling the most Difficult Diseases: Genetics and Genomics Open New Strategies to Fight Vector-Borne Diseases' (2001) 2:10 *EMBO Reports* 875.

Bubela, T.M. and T. Caulfield, 'Does the Print Media Hype Genetic Research?: A Comparison of Newspaper Stories and Peer Reviewed Research Papers' (2004) 170:9 *Canadian Medical Association Journal* 1399.

Buchanan A., 'Human Rights and the legitimacy of the international order' (2008), 14 *Legal Theory*, 39.

Buchanan, A., 'The Internal Legitimacy of Humanitarian Intervention' (1999) 7 *Journal of Political Philosophy* 71.

Buchanan, A.,'Rawls's Law of Peoples: Rules for a Vanished Westphalian World' (July 2000) *Ethics* 697.

Burch, R.K., P.J.D. Smith and W.P. Wheatley, 'Divergent Incentives to Protect Intellectual Property: A Political Economy Analysis of North–South Welfare' (2000) 3 *Journal of World Intellectual Property* 169.

Burk, D.L. and M.A. Lemley, 'Policy Levers in Patent Law' (2003) 89 *Va. L. Rev.* 1575.

Burke, W. et al., 'Recommendations for Follow-up Care of Individuals with an Inherited Predisposition to Cancer' (1997) 277:12 *Journal of the American Medical Association* 997.

Cahill, L.S., 'Genetics, Commodification, and Social Justice in the Globalisation Era' (2001) 11:3 *Kennedy Institute of Ethics Journal* 221.

Calland, R. and M. Taylor, 'Parliament and the Socio-Economic Imperative – What is the Role of the National Legislature?' (1997) 1 *Law, Democracy and Development* 193.

Caney, S., 'Cosmopolitan Justice and Equalising Opportunities' (2001) 32:1/2 *Metaphilosophy* 113.

Carlton, J.M. et al., 'Genome Sequence and Comparative Analysis of the Model Rodent Malaria Parasite Plasmodium Yoelii Yoelii' (2002) 419 *Nature* 512.

Carrasco, E. and M.A. Kose, 'Income Distribution and the Bretton-Woods Institutions: Promoting an Enabling Environment for Social Development' (1996) 6 *Transnat'l L. & Contemp. Probs.* 1.

Carroll, Sean B., 'Genetics and the Making of *Homo sapiens*' (2003) 422 *Nature* 849.

Caulfield, B.A., 'Why we Hate Gene Patents' (2003) *American Lawyer* 51.

Caulfield, T.A., 'Chaoulli v Quebec (Attorney General): The Supreme Court of Canada Deals a Blow to Publicly Funded Health Care' (2006) *Health Law Perspectives*.

Caulfield, T.A., 'Sustainability and the Balancing of the Health Care and Innovation Agendas: The Commercialization of Genetic Research' (2003) 66 *Sask. L. Rev.* 629.

Caulfield, T.A. and E.R. Gold, 'Genetic Testing, Ethical Concerns, and the Role of Patent Law' (2000) 57 *Clinical Genetics* 370.

Caulfield, T.A., E.R. Gold and M.K. Cho, 'Patenting Human Genetic Material: Refocussing the Debate' (2000) 1 *Nature Reviews Genetics* 227.

Chapman, A.R., 'Approaching Intellectual Property as a Human Right: Obligations Related to Article 15 (1) (c)' (2001) 35:3 *Copyright Bulletin* 4.

Chapman, A.R., 'Conceptualizing The Right To Health: A Violations Approach' (1998) 65 *Tenn. L. Rev.* 389.

Chapman, A.R., 'The Human Rights Implications of Intellectual Property Protection' (2002) 5 *Journal of Int'l Eco. L.* 861.

Chapman, A.R.,'Monitoring Women's Right to Health Under the International Covenant on Economic, Social and Cultural Rights' (1995) 44 *Am. U. L. Rev.* 1157.

Chen S. and M. Ravallion, 'How Have the World's Poorest Fared since the Early 1980s?' (2004) *World Bank Research Observer* 153.

Chien, C., 'Cheap Drugs at What Price to Innovation? Does Compulsory Licensing of Pharmaceuticals Hurt Innovation?' (2003) 18 *Berkeley Tech. L. J.* 853.

Chimni, B.S., 'International Institutions Today, An Imperial Global State in the Making' (February 2004) 15 *Eur. J. Int'l L.* 1.

Cho, M.K. et al., 'Effects of Patents and Licenses on the Provision of Clinical Genetic Testing Services' (2003) 5:1 *Journal of Molecular Diagnostics* 3.

Choi, D.H, et al., 'Incidence of BRCA1 and BRCA2 Mutations in Young Korean Breast Cancer Patients' (2004) 22:9 *J Clin Oncol.* 638.

Christiano, T., 'Democracy and Distributive Justice' (1995) 37 *Arizona Law Review* 65.

Collins, F.S., M. Morgan and A. Patrinos, 'The Human Genome Project: Lessons from Large-Scale Biology' (2003) 300: 5617 *Science* 286.

Collins, F.S., M. Morgan and A. Patrinos et al., 'A Vision for the Future of Genomics Research' (24 April 2003) 422 *Nature* 835.

Cook-Deegan, R.M. and S.J. McCormack, 'Patent Secrecy and DNA' (2001) 293 *Science* 217.

Cox, R.W., 'Civil Society at the Turn of the Millennium: Prospects for an Alternative World Order' (1999) 25 *Rev. Int'l Stud.* 3.

Crawford, M.H., 'Anthropological Genetics in the 21st Century: Introduction' (2000) 72 *Human Biology* 3.

Crespi, R.S., 'Models of Intellectual Property' (2002) 20 *Trends in Biotechnology* 451.

Crespi, R.S., 'Patenting and Ethics – A Dubious Connection' (January 2003) 85 *J. Pat. & Trademark Off. Soc'y* 31.

Cruft, R., 'Human Rights and Positive Duties' (Spring 2005) 19:1 *Ethics and International Affairs* 29.

Cullet, P., 'Patent and medicines: the relationship between TRIPs and the human right to health', (2003) 79:1 *International Affairs* 139.

Culyer, A.J. and A. Wagstaff, 'Equity and Equality in Health and Health Care' (1993) 12 *Journal of Health Economics* 431.

Czub, K.A., 'Argentina's Emerging Standard of Intellectual Property Protection: A Case Study of Underlying Conflicts Between Developing Countries, TRIPS standards and the United States' (2001) 33 *Case W. Res. J. Int'l. L.* 191.

Daar, A.S. et al., 'Top Ten Biotechnologies For Improving Health In Developing Countries' (October 2002) 32 *Nature Genetics* 269.

Damstedt, B.G., 'Limiting Locke: A Natural Law Justification for the Fair Use Doctrine' (2003) 112:5 *Yale Law Journal* 1179.

Daniels, N., 'Equality of What: Welfare, Resources, or Capabilities?' (1990) 50 (supplement) *Philosophy and Phenomenological Research* 273.

Daniels, N., 'Health Care Needs and Distributive Justice' (1981) 10 *Philosophy and Public Affairs* 146.

Daniels, N., B. Kennedy and I. Kawachi, 'Justice is Good for Our Health: How greater economic equality would promote public health' (February/March 2000) 25 *Boston Review*.

Darrow, M. and T. Amparo, 'Power, Capture, and Conflict: A Call for Human Rights Accountability in Development Cooperation' (2005) 27:2 *Human Rights Quarterly* 471.

Davies, J.L., 'Property interests in Human Reproductive Material' (2001) *Genetic Law Monitor* 6.

Davis, M.C., 'Constitutionalism and Political Culture: The Debate over Human Rights and "Asian Values"' (1998) 11 *Harv. Hum. Rts. J.* 109.

De Cock, K.M., D. Mbori-Ngacha and E. Marum, 'Shadow on the Continent: Public Health and HIV/AIDS in Africa in the 21st Century' (2002) 360 *Lancet* 67.

DeGrazia, D., 'Common Morality, Coherence, and the Principles of Biomedical Ethics' (2003) 13:3 *Kennedy Institute Journal of Ethics* 219.

DeMasi, T.E. and J.D. Garretson, 'PERSPECTIVE: Willful Patent Infringement Law Needs Reform' (2003) 230 *New York Law Journal* 5.

Demsetz, H., 'Toward a Theory of Property Rights II: The Competition Between Private and Collective Ownership' (2002) 31 *J.Legal Stud.* S.653.

Dodson, M. and R. Williamson, 'Indigenous Peoples and the Morality of the Human Genome Diversity Project' (1999) 25 *Journal of Medical Ethics* 204.

Dohnal, J., 'Structural Adjustment Programs: A Violation of Rights' (1994) 1 *Austl. J. Hum. Rts.* 57.

Dollar, D., 'The Poor Like Globalization but Institutions and Policies are Needed to Deliver the Hoped for Results' (23 June 2003) *Yale Global*, online: http://yaleglobal.yale.edu/display.article?id=1934 (accessed 4 June 2006).

Dolmo, B.-C., 'Examining Global Access to Essential Pharmaceuticals in the Face of Patent Protection Rights: The South African Example' (2001) 7 *Buff. Hum. Rts. L. Rev.* 137.

Dommen, C., 'Raising Human Rights Concerns in the World Trade Organization: Actors, Processes and Possible Strategies' (2002) 24 *Human Rights Quarterly* 1.

Dowdeswell, E., A.S. Daar and P.A. Singer, 'Bridging the Genomics Divide' (2003) 9 *Global Governance* 1.

Dryzek, J.S., 'Transnational Democracy' (March 1999) 7:1 *Journal of Political Philosophy* 30.

Dunoff, J.L., 'The Death of the Trade Regime' (1999) 10 *Eur. J. Int'l L.* 733.

Dutfield, G., 'Turning Knowledge into Power: Intellectual Property and the World Trade System' (2005) 59:4 *Australian Journal of International Affairs* 533.

Ebersole, T.J., M.C. Guthrie and J.A.Goldstein, 'Patent Pools as a Solution to the Licensing Problems of Diagnostic Genetics' (January 2005) 17:1 *Intellectual Property & Technology Law Journal* 6.

Eckert. A.E., 'Obligations beyond national borders: international institutions and distributive justice' (2008), Volume 4, No.1, *Journal of Global Ethics*.

Eggesrston, L., 'Ontario Defies US Firm's Genetic Patent, Continues Cancer Screening' (19 February 2002) 166:4 *CMAJ* 494.

Eisenberg, R.S., 'Patents and the Progress of Science: Exclusive Rights and Experimental Use' (1989) 56 *U. Chi. L. Rev.* 1017.

Eisenberg, R.S., 'Public Research and Private Development: Patents and Technology Transfer in Government-Sponsored Research' (1996) 82 *Va. L. Rev.* 1663.

Eisenberg, R.S., 'Re-Examining the Role of Patents in Appropriating the Value of DNA Sequences' (2000) 49 *Emory L. J.* 783.

Endy, D., 'Foundation for Engineering Biology' (2005) 438 *Nature* 449.

Enserink, M., 'Patent Office May Raise the Bar on Gene Claims' (2000) 287 *Science* 1196.

Enserink, M., 'Two New Steps Towards a 'Better Mosquito' (2000) 293 *Science* 2370.

Epstein, R.A., 'Why is Health Care Special?' (1993) 40 *U. Kan. L. Rev.* 307.

Evans, T., 'International Human Rights Law as Power/Knowledge' (2005) 27:3 *Human Rights Quarterly* 1046.

Evans, T., 'Universal Human Rights: As Much Round and Round As Ever Onward' (2003) 7:4 *The International Journal of Human Rights* 155.

Evans, T. and E. Hancock, 'Doing Something Without Doing Anything: International Human Rights Law and the Challenge of Globalisation' (1998) 2:3 *Int J Hum Rights* 1998 1.

Evans, W.E. and J.A. Johnson, 'Pharmacogenomics: The Inherited Basis for Interindividual Differences in Drug Response' (2001) 2 *Annu. Rev. Genomics Hum. Genet.* 9.

Falk, R., 'The Global Promise of Social Movements: Explorations at the Edge of Time' (1987) 12 *Alternatives* 173.

Falk, R., 'Social Movements and World Politics' (1997) 23:3 *Millenium (Special Issue).*

Farmer, P. et al., 'Community-Based Approaches to HIV Treatment in Resource-Poor Settings' (2001) 358:9279 *Lancet* 404.

Fentiman, L.C., 'AIDS as a Chronic Illness: A Cautionary Tale for the End of the Twentieth Century' (1998) 61 *Alb. L. Rev*. 989.

Fidler, D.P., 'International Law and Global Public Health' (1999) 48 *Kansas Law Review* 1.

Finlay, A., 'Gene Patenting: Seeking Benefits for All' (2003) 82 *Reform* 52.

Fleck, L.M., 'Just Health Care (I): Is Beneficence Enough?' (1989) 10 *Theoretical Medicine* 167.

Fleck, L.M., 'Just Health Care (II): Is Equal*ity Too Much*?' (1989) 10 *Theoretical Medicine* 301.

Flood, C.M. and T. Sullivan, 'Supreme Disagreement: The Highest Court Affirms an Empty Right' (2005) 173:2 *CJMA* 142.

Flores, M.A. and C. Campbell, 'Reexamining the Research Exemption' (2005) 23:5 *Nature Biotechnology* 659.

Forst, R., 'Toward a Critical Theory of Transnational Justice' (2001) 32:1/2 *Metaphilosophy* 26.

Forsythe, D., 'The United Nations, Human Rights, and Development' (1997) 19 *Human Rights Quarterly* 334.

Foster, M.W., 'Integrating Ethics and Science in the International HapMap Project' (2004) 5:6 *Nature Reviews Genetics* 467.

Frazier, M.E. et al., 'Realizing the Potential of the Genome Revolution: The Genomes to Life Program' (2003) 300: 5617 *Science* 290.

Gabel, P. and D. Kennedy, 'Roll Over Beethoven' (1984) 36 *Stan. L. Rev*. 1.

Garcia, F.J., 'The Universal Declaration of Human Rights at 50 and the Challenge of Global Markets: Trading Away the Human Rights Principle' (1999) 25 *Brook. J. Int'l L.* 51.

Gardner, M.J. et al., 'Genome Sequence of the Human Malaria Parasite Plasmodium Falciparm' (2002) 419 *Nature* 498.

Gathii, J.T., 'Construing Intellectual Property Rights and Competition Policy Consistency with Facilitating Access to Affordable AIDS Drugs to Low-End Consumers' (2001) 53 *Fla. L. Rev.* 727.

Gathii, J.T., 'Rights, Patents, Markets and the Global Aids Pandemic' (2002) 14 *Fla. J. Int'l L.* 261.

Gerhart, P., 'Reflections: Beyond Compliance Theory – TRIPS as a Substantive Issue' (2000) 32:3 *Case Western Reserve Journal of International Law* 357.

Gewirth, A., 'Are All Rights Positive?' (2001) 30:3 *Philosophy and Public Affairs* 321.

Gibson. D.G. et al., 'Complete Chemical Synthesis, Assembly, and Cloning of a Mycoplasma genitalium Genome' (29 February 2008) 319:5867 *Science* 1215.

Gill, S., 'Globalization, Market Civilization and Disciplinary Liberalism' (1995) 24:3 *Millennium: Journal of International Studies*.

Gill, S., 'Market Civilisation and Disciplinary Neoliberalism' (1995) 24: 3 *Millennium* 412.

Gold, E.R. and D.K. Lam, 'Balancing Trade in Patents: Public Non-Commercial Use and Compulsory Licensing' (2003) 6:1 *The Journal of World Intellectual Property* 5.

Gold, E.R. and D.K. Lam, 'Finding Common Cause in the Patent Debate' (2000) 18 *Nature Biotechnology* 1217.

Gold, E.R. and D.K. Lam, 'The Reach of Patent Law and Institutional Competence' (2003-2004) 1 *UOLTJ* 263.

Gold, E.R. and D.K. Lam,'From Theory to Practice: Health Care and the Patent System' (2003) Health Law Journal, Special Edition 2003: *Precedent & Innovation: Health Law in the 21st Century.*

Gold, E.R., D.K. Lam and A. Gallochat, 'The European Directive: Past and Prologue' (2001) 7:3 *European Law Journal* 331.

Gold, E.R. et al., 'Needed: Models of Biotechnology Intellectual Property' (2002) 20:8 *Trends in Biotechnology* 327.

Gold, E.R. et al., 'The Unexamined Assumptions of Intellectual Property: Adopting an Evaluative Approach to Patenting Biotechnology Innovation' (2004) *Public Affairs Quarterly* 299.

Gonzalez MacDowell, E., 'Juridical Action for the Protection of Collective Rights and its Legal Impact: a Case Study' (2002) 30:4 *The Journal of Law, Medicine & Ethics* 644.

Gorski, B. et al., 'A High Proportion of Founder BRCA1 Mutations in Polish Breast Cancer Families' (2004) 110:5 *Int J Cancer*. 683.

Gould, S.J., 'The Moral State of Tahiti – and of Darwin' (1991) 10 *Natural History* 12.

Grabowski, H., 'Patent, Innovation and Access to New Pharmaceuticals' (2002) *J. of Int'l. Eco L.*

Grady, M.F. and J.I. Alexander, 'Patent Law and Rent Dissipation' (1992) 78 *Va. L. Rev*. 305.

Greely, H.T., 'Legal, Ethical, and Social Issues in Human Genome Research' (1998) 27 *Annual Review of Anthropology* 473.

Griffin, J., 'First Steps in an Account of Human Rights' (2001) 9:3 *European Journal of Philosophy* 306.

Griffin, J., 'Discrepancies Between the Best Philosophical Account of Human Rights and the International Law of Human Rights' (2001) 101 *Proceedings of the Aristotelian Society* 1.

Guay, L.A. et al., 'Intrapartum and Neonatal Single-Dose Nevirapine Compared with Zidovudine for Prevention of Mother-to-Child Transmission of HIV-1 in Kampala, Uganda: HIVNET 012 Randomised Trial' (1999) 354 *Lancet* 795.

Guttmacher, A.E. and F.S. Collins, 'Genomic Medicine – a Primer' (2002) 347:19 *New England Journal of Medicine* 1512.

Hardin, G., 'The Tragedy of the Commons' (1968) 162 *Science* 1243.

Harries, A.D. et al., 'Preventing Antiretroviral Anarchy in Sub-Saharan Africa' (2001) 358: 9279 *Lancet* 410.

Harris, P.G. and P. Siplon, 'International Obligation and Human Health: Evolving Policy Responses to HIV/AIDS' (2001) 15:2 *Ethics and International Affairs* 1.

Hausman, D.M., Y. Asada and T. Hedemann, 'Health Inequalities and Why They Matter' (2002) 10 *Health Care Analysis* 177.

Haysom, N., 'Constitutionalism, Majoritarian Democracy and Socio-Economic Rights' (1992) 8 *SAJHR* 451.

Helfer, L.R., 'Regime Shifting: The Trips Agreement and New Dynamics of International Intellectual Property Lawmaking' (2004) 29 *Yale J. Int'l L.* 1.

Heller, M.A., 'The Tragedy of the Anticommons: Property in the Transition from Marx to Markets' (1998) 11 *Harv. L. Rev.* 621.

Heller, M.A. and R.S. Eisenberg, 'Can Patents Deter Innovation? The Anticommons in Biomedical Research' (1998) 280 *Science* 698.

Hendrie, H.C. et al., 'Alzheimer's Disease, Genes, and Environment: the Value of International Studies' (February 2004) 49:2 *Can J Psychiatry* 92.

Henkin, L., 'Human Rights and State "Sovereignty"' (1995) 25:1/2 *Georgia Journal of International and Comparative Law* 131.

Henry, M.R. et al., 'DNA Patenting and Licensing' (2003) 297 *Science* 1279.

Heredero, L., 'Comprehensive National Genetic Program in a Developing Country – Cuba' (1992) 28:3 *Birth Defects Original Article Series* 52.

Hessler, K., 'Resolving Interpretive Conflicts in International Human Rights Law' (2005) 13:1 *The Journal of Political Philosophy* 29.

Hettinger, E., 'Justifying Intellectual Property' (1989) 18 *Philosophy & Public Affairs* 32.

Hinojosa, J.P., 'The Human Genome, Property of All: Opportunities Under the ALRC Inquiry into Gene Patenting and Human Health' (2004) 26 *Sydney Law Review* 447.

Hinsch, W., 'Global Distributive Justice' (2001) 32:1/2 *Metaphilosophy* 58.

Hirschel, B. and P. Francioli, 'Progress and Problems in the Fight Against AIDS' (1998) 338 *New Eng. J. Med.* 906.

Holman, M.A. and S.R. Munzer, 'Intellectual Property Rights in Genes and Gene Fragments: A Registration Solution for Expressed Sequence Tags' (2000) 85 *Iowa L. Rev.* 735.

Holt, R.A. et al., 'The Genome Sequence of the Malaria Mosquito Anopheles Gambiae' (2002) 298 *Science* 129.

Holtzman, N.A. and T.M. Marteau, 'Will Genetics Revolutionize Medicine' (2000) 343:2 *The New England Journal of Medicine* 141.

Hudson, J., 'What Kinds of People Should we Create?' (2000) 17:2 *J Appl Philos* 131.

HUGO Ethics Committee, 'HUGO Urges Genetics Benefit-Sharing' (2000) 3 *Community Genet* 88.

Hugues, J., 'The Philosophy of Intellectual Property' (1988) 77:13 *Geo. L. J.* 287.

International Human Genome Sequencing Consortium, 'Initial Sequencing and Analysis of the Human Genome', (2001) 409 *Nature* 860.

Jackman, M., 'The Protection of Welfare Rights Under the Charter' (1988) 20 *Ottawa L. Rev.* 257.

Jackson, K.T., 'A Cosmopolitan Court for Transnational Corporate Wrongdoing: Why its Time has Come' (1998) *J. Bus Ethics* 758.

Jacobson, P.D. and S. Soliman, 'Co-opting the Health and Human Rights Movement' (Winter 2002) 30:4 *The Journal of Law, Medicine & Ethics* 705.

Juengst, E.T., 'FACE Facts: Why Human Genetics Will Always Provoke Bioethics' (2004) 32:2 *The Journal of Law, Medicine & Ethics* 26.

Johri, M. and C. Barry, 'Health and Global Justice' (2002) 16:2 *Health and International Affairs* 33.

Jomaa, H. et al., 'Inhibitors of the Nonmevalonate Pathway of Isoprenoid Biosynthesis as Antimalarial Drugs' (1999) 285 *Science* 1573.

Joseph, S., 'Pharmaceutical Corporations and Access to Drugs: The Fourth Wave of Corporate Human Rights Scrutiny' (2003) 25 *Human Rights Quarterly* 425.

Joyner, C., 'Legal Implications of the Concept of the Common Heritage of Mankind' (1986) 35:1 *International and Comparative Law Quarterly* 190.

Juma, C. and L. Yee-Cheong, 'Reinventing Global Health: the Role of Science, Technology and Innovation' (2005) 365 *Lancet* 1105.

Kaiser, J., 'Biobanks: Population Databases Boom, from Iceland to the U.S.' (2002) 298:5596 *Science* 1158.

Kato, N. 'Genetic Analysis in Human Hypertension' (2002) 25:3 *Hypertens Res.* 319.

Kennedy, D., 'The International Human Rights Movement: Part of the Problem?' (2002) 15 *Harvard Human Rights Journal* 101.

Kent, H., 'BC Sidesteps Patent Claim, Transfers *BRCA* Gene Testing to Ontario' (2003) 168:2 *CMAJ* 211.

Khan, R., 'The Anti-Globalization Protests: Side-Show of Global Governance, or Law-Making on the Streets?' (2001) 61 *Heidelberg Journal of International Law* 323.

Khouri, M.J., 'Genetics and Genomics in Practice: The Continuum From Genetic Disease to Genetic Information in Health and Disease' (2003) 5:4 *Genetic Medicine* 261.

Khoury, R. et al., 'Population Screening in the Age of Genomic Medicine' (2003) 348 *New England Journal of Medicine* 50.

Kieff, F.S., 'IP Transactions: On The Theory & Practice of Commercializing Innovation' (2005) 42:3 *Houston Law Review* 727.

Kieff, F.S., 'Property Rights and Property Rules for Commercializing Inventions' (2001) 85 *Minn L. Rev*. 697.

Kieff, F.S., 'The Case for Registering Patents and the Law and Economics of Present Patent-Obtaining Rules' (2003) 45 *B.C. L. Rev*. 55.

Kieff, F.S., 'Facilitating Scientific Research: Intellectual Property Rights and the Norms of Science—A Response to Rai and Eisenberg' (2000) 95 *NW. U. L. Rev*. 691.

Kirby, M., 'Genomics and Democracy, A Global Challenge' (2003) 31:1 *UWA Law Review* 1.

Kirby, M., 'The Right to Health Fifty Years on: Still Sceptical?' (1999) 4:1 *Health Hum Rights* 6.

Klaaren, J., 'A Second Look at the South African Human Rights Commission, Access to Information, and the Promotion of Socioeconomic Rights' (2005) 27:2 *Human Rights Quarterly* 539.

Knoppers, B.M., 'Status, Sale and Patenting of Human Genetic Material: An International Survey' (1999) 22 *Nature Genetics* 23.

Koskenniemi, M., 'The Future of Statehood' (1991) 32 *Harv. J. Int'l L*. 397.

Kothari, R., 'Globalization: A World Adrift' (1997) 22 *Alternatives* 227.

Kowalski, T., 'International Patent Rights and Biotechnology: Should the United States promote Technology Transfer of Developing Countries?' (2002) 25 *Loy.L.A. Int'l. & Comp. Law Rev*. 41.

Kupler, A., 'Debate: Global Poverty Relief, More than Charity: Cosmopolitan Alternative to the Singer Solution' (2002) 16:1 *Ethics and International Affairs* 107.

Kupler, A., 'Rawlsian Global Justice, Beyond the Law of Peoples to a Cosmopolitan law of Persons' (October 2000) 28:5 *Political Theory* 640.

Lahat, H. et al., 'A Missense Mutation in a Highly Conserved Region of *CASQ2* is Associated with Autosomal Recessive Catecholamine-Induced Polymorphic Ventricular Tachycardia in Bedouin Families from Israel' (2001) 69 *American J Human Genetics* 1378.

Lampis, R. et al., 'The Distribution of HLA Class II Haplotypes Reveals that the Sardinian Population is Genetically Differentiated from the Other Caucasian Populations' (2000) 56 *Tissue Antigens* 515.

Lange, D., 'Recognising the Public Domain' (1981) 44:4 *Law and Contemporary Problems* 147.

Langlois, A.J., 'Human Rights: The Globalization and Fragmentation of Moral Discourse' (2002) 28 *Rev. Int'l Stud*. 479.

Lawson, C., 'Patenting Genetic Diversity, Old Rules May be Restricting the

Exploration of a New Technology' (1999) 6 *Journal of Law and Medicine* 373.

Leary, V.A., 'The Right to Health in International Human Rights Law' (1994) 1 *Health and Human Rights* 24.

Lemke, T., 'Beyond Genetic Discrimination. Problems and Perspectives of a Contested Notion' (2005) 1:3 *Genomics, Society and Policy* 22.

Leskien, D. and M. Flitner, 'Intellectual Property Rights and Plant Genetic Resources: Options for a Sui Generis System' (1997) 6 *Issues in Genetic Resources*.

Levinson, J., 'The International Financial System: A Flawed Architecture' (1999) 23:1 *The Fletcher Forum on World Affairs* 1.

Liebenberg, S., 'The Right to Social Assistance: The Implications of Grootboom' (2001) 1:7 *S. Afr. J. of Hum. Rts.* 232.

Liivak, O., 'The Forgotten Originality Requirement: A Constitutional Hurdle for Gene Patents' (2005) 85:4 *Journal of Patent and Trademark Office Society* 261.

Link, Bruce G. et al., 'Social Epidemiology and the Fundamental Cause Concept: on the Structuring of Effective Cancer Screens by Socioeconomic Status' (1998) 76 *Milbank Quarterly* 375.

Lipinski, T.A. and J. Britz, 'Rethinking the Ownership of Information in the 21st Century: Ethical Implications' (2000) 2:1 *Ethics and Information Technology* 49.

Livingston, D.M. and R. Shivdasani, 'Towards Mechanism-Based Cancer Care' (2001) 285 *Journal of the American Medical Association* 588.

Loewenson, R., 'Essential Drugs in Southern Africa Need Protection from Public Health Safeguards under TRIPs' (2000) 4:7 *Bridges Between Trade & Sustainable Dev.* 3.

London, L., 'Human rights and Public Health: Dichotomies or Synergies in Developing Countries? Examining the Case of HIV in South Africa' (2002) 30:4 *The Journal of Law, Medicine & Ethics* 677.

Lynch, J.W., G.A. Caplan and J.T. Salonen, 'Why do Poor People Behave Poorly? Variation of Adult Health Behaviours and Characteristics by Stages of the Socioeconomic Life Course' (1997) 44:6 *Social Science and Medicine* 809.

Mackay, E., 'Economic Incentives in Markets For Information And Innovation' (1990) 13 *Harvard Journal of Law & Public Policy* 867.

MacLean, M. et al., 'Making Malaria Research Bite' (1997) 3:1 *Nature Medicine* 14.

Maioni, A. and C. Manfredi, 'When the Charter Trumps Health Care – A Collision of Canadian Icons' (2005) *Policy Options* 52.

Malander, S., 'One in 10 Ovarian Cancer Patients Carry Germ Line BRCA1 or BRCA2 Mutations: Results of a Prospective Study in Southern Sweden' (2004) 40:3 *Eur J Cancer* 422.

Mandela, N., *Address at the 53rd United Nations General Assembly*, New York, 21 September 1998 quoted in M. Heywood and D. Altman, 'Confronting AIDS: Human Rights, Law and Social Transformation' (2000) 5:1 *Health and Human Rights* 149.

Mansur, L., 'Gene Discovery, Ownership and Access for Developing Countries in the Era of Molecular Genetics' (2002) 5:1. *Electronic Journal of Biotechnology* online: http://www.ejbiotechnology.info/content/vol5/issue1/issues/05/ (accessed 16 May 2006).

Mapulanga-Hulston, J.K., 'Examining the Justiciability of Economic, Social and Cultural Rights' (2002) 6:4 *The International Journal of Human Rights* 29.

Marchand, S., D. Wilker and B. Landesman, 'Class, Health and Justice' (1998) 76:3 *Milbank Quarterly* 449.

Marks, S.P., 'Tying Prometheus Down: The International Law of Human Genetic Manipulation' (2002) 3 *Chi. J. Int'l L.* 115

Marshall P. and B. Koenig, 'Accounting for Culture in a Globalised Bioethics' (2004) 32 *Journal of Law, Medicine and Ethics* 252.

Martinez P.A. et al., 'Simple and Rapid Detection of the Newly Described Mutations in the HLA-H Gene' (1997) 89 *Blood* 1835.

Martone, M., 'The Ethics of the Economics of Patenting the Human Genome' (1998) 17 *J. Bus. Ethics* 1679.

Maskus, K., 'Normative Concerns in the International Protection of Intellectual Property Rights' (1991) 14 *World Economy* 403.

Mason, K. and G. Laurie, 'Consent or Property? Dealing with the Body and its Parts in the Shadow of Bristol and Alder Hey' (2001) 64:5 *The Modern Law Review* 725.

May, C., 'Cosmopolitan Legalism Meets Thin Community: Problems in the Global Governance of IP' (2004) *Government and Opposition* 393.

May, C., 'Unacceptable Costs: The Consequences of Making Knowledge Property in a Global Society' (2002) 16:2 *Global Society* 123.

McCall Smith, A., 'Property, Dignity and the Human Body' (1994) 2:3 *Hume Papers on Public Policy* 29.

McCalman, P., 'Reaping What You Sow: An Empirical Analysis of International Patent Harmonization' (2001) 55 *Journal of International Economics* 161.

McCorquodale, R., 'Self-Determination: A Human Rights Approach' (1994) 43 *Int'l & Comp. L.Q.* 857.

McCorquodale, R. and R. Fairbrother, 'Globalization and Human Rights' (1999) 21:3 *Human Rights Quarterly* 735.

Mckenna, E.M., 'The Mandatory Testing of Newborns for HIV: Too Much, Too Little, Too Late' (1997) 13 *N.Y.L. Sch. J. Hum. Rts.* 307.

McGwire, M., 'The Paradigm that Lost its Way' (2001) 77:4 *International Affairs* 777.

Merz, J.F. et al., 'Diagnostic Testing Fails the Test: The Pitfalls of Patents Are Illustrated by the Case of Haemochromatosis' (2002) 415 Nature 577.

Merz, J.F. and M.K. Cho, 'What are Gene Patents and Why are People Worried about Them?' (2005) 8:4 *Community Genetics* 203

Merz, J.F., G.E. McGee and P. Sankar, 'Iceland Inc.?: On the ethics of commercial population genomics' (2004) 58:6 *Soc Sci Med.* 1201.

Merz, J.F. 'Disease Gene Patents: Overcoming Unethical Constraints on Clinical Laboratory Medicine' (1999) 45:3 *Clinical Chemistry* 324.

Meyer, W.H., 'Human Rights and MNCs: Theory Versus Quantitative Analysis' (1996) 18 *Human Rights Quarterly* 368.

Mira, M.T., 'Chromosome 6q25 is Linked to Susceptibility to Leprosy in a Vietnamese Population' (2003) 33 *Nature Genetics* 412.

Monshipouri, M., C.E. Welch and E.T. Kennedy, 'Multinational Corporations and the Ethics of Global Responsibility: Problems and Possibilities' (2003) 25:4 *Human Rights Quarterly* 965.

Monshipouri, M., C.E. Welch and E.T. Kennedy, 'Shifting the Paradigm' (2001) 77:1 *International Affairs* 1.

Moellendorf, D., 'Constructing the Law of Peoples' (1996) *Pacific Philosophical Quarterly* 77.

Moellendorf, D., 'The World Trade Organization and Egalitarian Justice' (2005) 36:1/2 *Metaphilosophy* 145.

Mowzoon, M., 'Access Versus Incentive: Balancing Policies in Genetic Patents' (2003) 35 *Ariz. St. L.J.* 1077.

Muchlinski, P.T., 'Human Rights and Multinationals: Is There a Problem?' (2001) 77:1 *International Affairs* 31.

Mukherjee, J., 'HIV-1 Care in Resource-Poor Settings: A View from Haiti' (2003) 362 *Lancet* 994.

Nance, D.A., 'Foreword: Owning Ideas' (1990) 13 *Harvard Journal of Law & Public Policy* 757.

National Institute of Health, 'Best Practices for the Licensing of Genomic Inventions' (2005) 70 *Federal Register* 18413.

Nelkin, D. and L. Andrews, 'Homo Economicus: Commercialization of Body Tissue in the Age of Biotechnology' (1998) 28:5 *Hastings Center Report* 30.

Nelsen, L., 'The Role of University Technology Transfer Operations in Assuring Access to Medicines and Vaccines in Developing Countries' (2003) III:2 *Yale Journal of Health Policy Law and Ethics* 301.

Nerozzi, M.M., 'The Battle Over Life-Saving Pharmaceuticals: Are Developing Countries Being "TRIPped" by Developed Countries' (2002) 47 *Vill. L. Rev.* 605.

Newson, A. and R. Williamson, 'Should we Undertake Genetic Research on Intelligence?' (1999) 13:3/4 *Bioethics* 327.

Ngwena, C., 'Access to Health Care a Fundamental Right: The Scope and

Limits of Section 27 of the Constitution' (2000) 25 *Journal for Juridical Sciences* 1.

Nussbaum, M., 'Human Functioning and Social Justice: in Defense of Aristotelian Essentialism' (1992) 20 *Political Theory* 222.

Nwabueze, R.N., 'Ethnopharmacology, Patents and the Politics of Plants' Genetic Resources' (2003) 11 *Cardozo J. Int'l & Comp. L.* 585.

Nzomo, M., 'The Political Economy of the African Crisis: Gender Impacts and Responses' (1996) 51 *Int'l J.* 78.

Ocampo, J.A., 'Rethinking the Development Agenda' (2002) 26:3 *Cambridge Journal of Economics* 393.

Ocampo, J.A., 'Globalization, Development and Democracy' (2005) 5:3 Items and Issues, online on the website of *Yale Global* http://yaleglobal.yale.edu/about/pdfs/ocampo.pdf (accessed 30 May 2006).

Oddi, A.S., 'TRIPS: Natural Rights and a Polite Form of Economic Imperialism' (May 1996) *Vanderbilt Journal of Transnational Law* 29.

Oddi, A.S., 'Un-Unified Economic Theories of Patents – the Not-Quite-Holy Grail' (1996) 71 *Notre Dame L. Rev.* 267.

Ogbu, O., 'African Science and Technology Day: Has Africa Surrendered in the Field of Science and Technology?' (2002) *Science and Development in Africa.*

O'Keefe, E. and A. Scott-Samuel, 'Human Rights and Wrongs: Could Health Impact Assessment Help?' (2002) 30:4 *The Journal of Law, Medicine & Ethics* 734.

Oloka-Onyango, J., 'Beyond the Rhetoric: Reinvigorating the Struggle for Economic and Social Rights in Africa' (1995) 26 *Cal. W. Int'l L.J.* 1.

Oloka-Onyango, J., 'Reinforcing Marginalised Rights in an Age of Globalisation: International Mechanisms, Non-State Actors and the Struggle for Peoples' Rights in Africa' (2003) 18 *Am. U. Int'l L. Rev.* 851.

O'Neill, O., 'Agents of Justice' (2001) 32:1/2 *Metaphilosophy* 180.

O'Neill, O., 'Public Health or Clinical Ethics: Thinking beyond Borders' (2002) 16:2 *Ethics & International Affairs* 35.

Orme, I.M., D.N. McMurray and J.T. Belisle, 'Tuberculosis Vaccine Development: Recent Progress' (2001) 9 *Trends in Microbiology* 115.

Osinbajo, Y. and O. Ajayi, 'Human Rights and Economic Development in Developing Countries' (1994) 28 *Int'l Law* 727.

Ostergard, Jr., R. L., 'Intellectual Property: A Universal Human Right?' (1999) 21:1 *Human Rights Quarterly* 156.

Ottaway, M., 'Reluctant Missionaries' (2001) *Policy* 44.

Palmer, T.G., 'Are Patents and Copyrights Morally Justified? The Philosophy of Property Rights and Ideal Objects' (1990) 13 *Harvard Journal of Law & Public Policy* 860.

Palmeter, D. and P.C. Mavriodis, 'The WTO Legal System: Sources of Law' (1998) 92 *Am. J. Int'l L.* 398.

Paradise, J., L.B. Andrews and T. Holbrook, 'Patents on Human Genes: an Analysis of Scope and Claims' (2005) 307:5715 *Science* 1566.

Participant in the 2001 Conference on Ethics Aspects of Research in Developing Countries, 'Fair Benefits for Research in Developing Countries' (2002) 298 *Science* 2133.

Peltonen, L. and V.A. Mckusick, 'Dissecting Human Disease in the Postgenomic Era' (2001) 291:5507 *Science* 1224.

Peters, D.H. et al., 'Reducing the Impact of Poverty on Health and Human Development: Scientific Approaches', *Annals of the New-York Academy of Sciences* (2008) Vol. 1136.

Petchesky, R.P., 'Rights and Needs: Rethinking the Connections in Debates over Reproductive and Sexual Rights' (2000) 4:2 *Health and Human Rights* 17.

Petersmann, E.-U., 'How to Constitutionalize International Law and Foreign Policy for the Benefit of Civil Society?' (1998) 20 *Michigan Journal of International Law* 1.

Petersmann, E.-U., 'The WTO Constitution and Human Rights' (2000) 3:1 *Journal of International Economic Law* 19.

Petrou, M. and B. Modell, 'Prenatal Screening for Haemoglobin Disorders' (1995) 15 *Prenat. Diagn* 1275.

Pevnick, R., 'Political Coercion and the scope of Distributive Justice' (2008) 56 *Political Studies*.

Pleiss, Jürgen, 'The Promise of Synthetic Biology', (2006) 73 *Appl. Microbiol. Biotechnol.* 735.

Pogge, T.W., 'Recognized and Violated by International Law: The Human Rights of the Global Poor' (2005) 18:4 *Leiden Journal of International Law* 717.

Pogge, T.W., 'Human Rights and Global Health: A Research Program' (2005) 36:1/2 *Metaphilosophy* 182.

Pogge, T.W., 'Symposium on World Poverty and Human Rights' (2005) 19:1 *Ethics & International Affairs* 1.

Pogge, T.W., 'The Moral Demands of Global Justice' (2003) *Dissent* 37.

Pogge, T.W., 'Moral Universalism and Global Economic Justice' (2002) 1:1 *Politics, Philosophy and Economics* 29.

Pogge, T.W., 'Responsibilities for Poverty-Related Ill Health' (2002) 16:2 *Ethics and International Affairs* 71.

Pogge, T.W., 'Severe Poverty as a Violation of Negative Duties' (Spring 2005) 19:1 *Ethics and International Affairs* 55.

Pogge, T.W., 'Cosmopolitanism and Sovereignty' (October 1992) 103 *Ethics* 49.

Pogge, T.W., 'An Egalitarian Law of Peoples' (Summer 1994) 23 *Philosophy and Public Affairs* 211.

Pogge, T.W., 'Eradicating Systematic Poverty: Brief for a Global Resources Dividend' (2001) 2 *Journal of Human Development* 59.

Prager, F.D., 'The Early Growth and Influence of Intellectual Property' (1952) 34 *J. Pat. Off. Soc'y* 106.

Pressman, L. et al., 'The Licensing of DNA Patents by US Academic Institutions: an Empirical Survey' (2006) 24 *Nature Biotechnology* 21.

Primo Braga, C. and C. Fink, 'The Relationship Between Intellectual Property Rights and Foreign Direct Investment' (1998–1999) 9 *Duke J. Comp. & Int'l L.* 163.

Rai, A.K., 'Evolving Scientific Norms and Intellectual Property Rights: A Reply to Kieff' (2001) 95:2 *Northwestern University Law Review* 707.

Rai, A.K., 'Regulating Scientific Research: Rights and the Norms of Science in Biotechnology Research' (1999) 94 *NW. U. L. Rev.* 77.

Rajagopal, B., 'From Resistance to Renewal: The Third World, Social Movement, and the Expansion of International Institutions' (2000) 41 *Harv. Int'l L. J.* 529.

Ramos, R.G. and K. Olden, 'Gene-Environment Interactions in the Development of Complex Disease Phenotypes' (March 2008) 5:1 *Int. J. Environ Res Public Health* 4.

Ransom, D., 'The Dictatorship of Debt' (October 1999) *World Press Rev* 6.

Ratner, S.R., 'Corporations and Human Rights: A Theory of Legal Responsibility' (2001) 111 *Yale L. J.* 461.

Raustiala, K., 'Compliance and Effectiveness in International Regulatory Co-operation' (Summer 2000) 32:3 *Case Western Reserve Journal of International Law* 387.

Reichman, J., 'From Free Traders to Fair Followers: Global Competition Under the TRIPS Agreement' (1997) 29 *N.Y.U.J. Int'l L. & Pol.* 11.

Reichman, J., 'Universal Minimum Standards of Intellectual Property Protection under the TRIPS Component of the WTO Agreement' (1996) 29 *Int'l L.* 345.

Resnik, D.B., 'Fair Drug Prices and Patent System' (2004) 12:2 *Health Care Analysis* 91.

Resnik, D.B., 'DNA Patents and Human Dignity' (2001) 29:2 *The Journal of Law, Medicine, and Ethics* 152.

Resnik, D.B., 'DNA Patents and Scientific Discovery and Innovation: Assessing Benefits and Risks' (2001) 7:1 *Science and Engineering Ethics* 29.

Rhodes, R., 'Genetic Links, Family Ties, and Social Bonds: Rights and Responsibilities in the Face of Genetic Knowledge' (1998) 23:1 *Journal of Medicine and Philosophy* 10.

Rice, C., 'Campaign 2000, 'Promoting the National Interest' (2000) 79 *Foreign Affairs* 45.

Rich, G. S., 'The Relation Between Patent Practices and the Anti-Monopoly Laws' (1942) 24 *J. Pat. Off. Soc'y* 85.

Richardson, J.L., 'Contending Liberalisms: Past and Present' (1997) 3 *Eur. J. of Int'l Rel.* 5.

Rimmer, M., 'The Freedom to Tinker: Patent Law and Experimental Use' (2005) 15:2 *Expert Opinion* 167.

Rittich, K., 'The Future of Law and Development: Second Generation Reforms and the Incorporation of the Social' (2004) 26 *Mich. J. Int'l L.* 199.

Rittich, K., 'Transformed Pursuits: The Quest for Equality in Globalized Markets' (2002) 13 *Harv. Hum. Rts. J.* 231.

Robertson, J.A., 'The $1000 Genome: Ethical and Legal Issues in Whole Genome Sequencing of Individuals' (2003) 3:3 *The American Journal of Bioethics* W35.

Robertson, J.A., 'Critical Reflection, on the Politics of Need: Implications for Public Health' (1998) 47 *Social Science and Medicine* 1419.

Robertson, R., 'Measuring State Compliance with the Obligation to Devote the Maximum Available Resources to Realising Economic Social and Cultural Rights' (1994) 16 *Human Rights Quarterly* 693.

Robinson, K., 'False Hope or a Realizable Right? The Implementation of the Right to Shelter under the African National Congress' Proposed Bill of Rights for South Africa' (1993) 28 *Harvard Civil Rights–Civil Liberties Law Review* 505.

Roos, D.S., 'Bioinformatics – Trying to Swim in a Sea of Data' (2001) 291 *Science* 1260.

Roses, A.D., 'Pharmacogenetics and the Practice of Medicine' (2000) 405 *Nature* 857.

Roth, K., 'Defending Economic, Social and Cultural Rights: Practical Issues Faced by an International Human Rights Organization' (2004) 26 *Human Rights Quarterly* 63.

Sadasivam, B., 'The Impact of Structural Adjustment on Women: A Governance and Human Rights Agenda' (1997) 19 *Hum. Rts. Q.* 630.

Sakakibara, M. and L. Branstetter, 'Do Stronger Patents Induce More Innovation? Evidence From the 1988 Japanese Patent Law Reforms' (2001) 32 *RAND Journal of Economics* 77.

Sarkin, J., 'Health and Human Rights in Post-Apartheid South Africa' (1999) 89 *South African Medical Journal* 1259.

Scheffler, S., 'Conceptions of Cosmopolitanism' (1999) 11 *Utilitas* 255.

Scheffler, S., 'Relationships and Responsibilities' (1997) *Philosophy and Public Affairs* 26.

Scherer, F.M., 'Le Système de Brevet et l'Innovation dans le Secteur Pharmaceutique/ The Patent System and Innovation in Pharmaceuticals' (2000) 1 Revue Internationale de Droit Economique 110.

Schissel, A., J.F. Merz and M.K. Cho,'Survey Confirms Fears About Licensing of Genetic Tests' (1999) 402 *Nature* 118.

Scott, C. and P. Macklem, 'Constitutional Ropes of Sand or Justiciable Guarantees?: Social Rights in a New South African Constitution' (1992) 141 *U. Pa. L. Rev.* 1.

Séguin, B. et al., 'Human Genomic Variation Initiatives in Emerging Economies and Developing Countries' (October 2008) *Nature Reviews, Genetics – Genomic Medicine in Developing Countries*, online www.nature.com/nrg/supplements/genomicmedicine (accessed 26 April 2009).

Ségin, B. et al., 'Genomic Medicine and Developing Countries: Creating a Room of their Own' (June 2008) 9 *Nature Review Genetics* 487.

Sell, S. and C. May, 'Moments in Law: Contestation and Settlement in the History of Intellectual Property' (2001) 8:3 *Review of International Political Economy* 467.

Sen, A., 'Consequential Evaluation and Practical Reason' (2000) 17:9 *The Journal of Philosophy* 478.

Service, R., 'Genetics and Medicine: Recruiting Genes, Proteins for a Revolution in Diagnostics' (2003) 300:5617 *Science* 236.

Sheremeta, L. and B.M. Knoppers, 'Beyond the Rhetoric: Population Genetics and Benefit-Sharing' (2003) 11 *Health Law Journal* 89.

Shickle, D., 'Are "Genetic Enhancements" Really Enhancements?' (2000) 9:3 *Camb Q Healthc Ethics* 342.

Shifman, S. and A. Darvasi,'The Value of Isolated Populations' (2001) 28 *Nature Medicine* 309.

Shue, H., 'The Burden of Justice' (1983) 80 *Journal of Philosophy* 600.

Sidley, P., 'South African Court Battle Damages Drug Industry's Image' (2001) 322 *BMJ* 635.

Simm, K., 'Benefit-Sharing Regarding the Meaning and Limits of the Concept in Human Genetic Research' (2005) 1:2 *Genomics, Society and Policy* 29.

Singer, P.A. and A.S. Daar, 'Harnessing Genomics and Biotechnology to Improve Global Health Equity' (2001) 294 *Science* 87.

Skogly, S. and M. Gibney, 'Transnational Human Rights Obligations' (2002) 24:3 *Human Rights Quarterly* 781.

Smith, M.J., 'Population-Based Genetic Studies: Informed Consent and Confidentiality' (2001) 18:1 *Santa Clara Comput High Technol Law J.* 57.

Smith, P.J., 'Are Weak Patent Rights a Barrier to U.S. Exports?' (1999) 48 *Journal of International Economics* 151.

Smith, P.J., 'Patent Rights and Trade: Analysis of Biological Products, Medicinals and Botanicals, and Pharmaceuticals' (2002) 84 *Amer. J. Agr. Econ.* 495.

Sorenson, J. and J. Botkin (eds), 'Genetic Testing and the Family' (2003) 119C *American Journal of Medical Genetics Pan G Seminars in Medical Genetics* 1.

Spectar, M., 'The Fruit of The Human Genome Tree: Cautionary Tales about Technology, Investment, and the Heritage of Mankind' (2001) 23:1 *Loyola of Los Angeles International and Comparative Law Review* 1.

Spectar, M., 'Patent Necessity: Intellectual Property Dilemmas in the Biotech Dominant Treatment Equity for Developing Countries' (2001) 24 *Hous. J. Int'l L* 227.

Stammers, N., 'Social Movements and the Social Construction of Human Rights' (1999) 21:4 *Human Rights Quarterly* 980, at 991.

Steidlmeier, P., 'The Moral Legitimacy of Intellectual Property Claims: American Business and Developing Country Perspectives' (1993) 12:3 *Journal of Business Ethics* 162.

Stein, L., 'Genome Annotation: from Sequence to Biology' (2001) 2 *Nature Reviews Genetics* 493.

Streit, C. et al., 'CFTR gene: Molecular Analysis in Patients from South-Brazil' (2003) 78 *Molecular Genetics and Metabolism* 259.

Su, E., 'The Winners and the Losers: The Agreement on Trade-Related Aspects of Intellectual Property Rights and Its Effects on Developing Countries' (2000) 23 *Hous. J. Int'l L.* 169.

Tabor, A.-M., 'Recent Development: AIDS Crisis' (2001) 38 *Harv. J. on Legis.* 514.

Tan, K.-C., 'Kantian Ethics and Global Justice' (1997) 23:1 *Social Theory and Practice* 53.

Tesón, F.R., 'International Human Rights and Cultural Relativism' (1985) 25 *Va. J. Int'l L.* 869.

Tesón, F.R., 'Some Observations on John Rawls's "The Law of Peoples" ' (1994) 88 *American Society of International Law Proceedings* 18.

Tesón, F.R., 'The Rawlsian Theory of International Law' (1995) *Ethics and International Affairs* 9.

Thorsteinsdottir, H. et al., 'Genomics – a Global Public Good?' (2003) 361:9361 *The Lancet* 891.

Thurow, L.G., 'Needed: A New System of Intellectual Property Rights' (1997) *Harv. Bus. Rev.* 95.

Tobin, J., 'On Limiting the Domain of Inequality' (1970) 13:2 *The Journal of Law and Economics* 263.

Toebes, B., 'Towards an Improved Understanding of the International Right to Health' (1999) 21:3 *Human Right Quarterly* 662.

Tolchin B., 'Human Rights and the Requirement for International Medical Aid' (2008) 8:2 *Developing World Bioethics* 151.

Torres, M.A., 'The Human Right to Health, National Courts, and Access to HIV/AIDS Treatment: A Case Study from Venezuela' (2002) 3:1 *Chicago Journal of International Law* 105.

Trouiller, P. et al., 'Drug Development for Neglected Diseases: A Deficient Market and a Public-Health Policy Failure' (2002) 359:9324 *Lancet* 2188.

Tsien, J. et al., 'Genetic Enhancement of Learning and Memory in Mice' (1999) 401 *Nature* 63.

Tushnet, M., 'An Essay on Rights' (1984) 62 *Texas Law Review* 1363.

Varmus, H., 'Getting Ready for Gene-Based Medicine' (2002) 247 *N. Eng. J. Med.* 1526.

Venter, J.C. et al., 'The Sequence of the Human Genome' (2001) 291:5507 *Science* 1304.

Verma, S.P., 'Malaria Genome Project and its Impact on Disease' (2003) 40:1/2 *Journal of Vector Borne Diseases* 9.

Wade, R., 'Japan, the World Bank, and the Art of Paradigm Maintenance: The East Asian Miracle in Political Perspective' (1996) 217 *New Left Rev*. 3.

Walsh, J.P., C. Cho and W.M. Cohen, 'Science and Law. View from the Bench: Patents and Material Transfers' (Sept. 23, 2005) 309:5743 *Science* 2002.

Warner, M.A., 'Globalization and Human Rights: An Economic Model' (1999) 25 *Brook. J. Int'l L.* 99.

Watal, J., 'Pharmaceutical Patents, Prices and Welfare Losses: Policy Options for India Under the WTO TRIPS Agreement' (2000) 23:5 *World Economy* 733.

Wikler, D., 'Personal and Social Responsibility for Health' (2002) 16:2 *Ethics and International Affairs* 47.

Williams, A.D., 'The Revisionist Difference Principle' (1995) 25:2 *Canadian Journal of Philosophy* 257.

Willison, D. and Macleod, S.M., 'Patenting of Genetic Material: are the Benefits to Society Being Realised?' (2006) 167:3 *Canadian Medical Association Journal* 259.

Winkelmann, B.R., 'Pharmacogenomics, Genetic Testing and Ethnic Variability: Tackling the Ethical Questions' (2003) 4:5 *Pharmacogenomics* 531.

Winston, M., 'NGO Strategies for Promoting Corporate Social Responsibility' (2002) 16 *Ethics & Int'l Aff.* 71.

Yamin, A.E., 'Protecting and Promoting the Right to Health in Latin America: Selected Experiences from the Field' (2000) 5:1 *Health and Human Rights* 117.

Index